Fernando Pérez-Rodríguez • Antonio Valero

Predictive Microbiology in Foods

 Springer

Fernando Pérez-Rodríguez
Dpt Bromatologia y Tec.
  de los Alimentos
University of Córdoba
Córdoba, Spain

Antonio Valero
Dpt Bromatologia y Tec.
  de los Alimentos
University of Córdoba
Córdoba, Spain

ISBN 978-1-4614-5519-6   ISBN 978-1-4614-5520-2 (eBook)
DOI 10.1007/978-1-4614-5520-2 100700 5227
Springer New York Heidelberg Dordrecht London

Library of Congress Control Number: 2012950350

Printed on acid-free paper

Springer is part of Springer Science+Business Media (www.springer.com)

SpringerBriefs in Food, Health, and Nutrition Series

Springer Briefs in Food, Health, and Nutrition present concise summaries of cutting edge research and practical applications across a wide range of topics related to the field of food science.

Editor-in-Chief

Richard W. Hartel, *Un*

Associate Editors

J. Peter Clark, *Consul*
John W. Finley, *Louis*
David Rodriguez-Laza
David Topping, *CSIR*

For further volumes:
http://www.springer.com

Contents

Chapter 1
Predictive Microbiology in Foods

Abstract Predictive microbiology in foods is a research area within food microbiology intended to provide mathematical models to predict microbial behavior in food environments. Although the first predictive models were dated at the beginning of the 20th century, its great development has occurred in the past decades as a result of computer software advances. In addition to the exhaustive knowledge on food microbiology, the predictive microbiology field is based on important mathematical and modeling concepts that should be previously introduced for predictive microbiology beginners. The different typology of predictive models allows predicting growth, inactivation, and probability of growth of bacteria in foods under different environmental conditions and considering additional factors such as the physiological state of cells or interaction with other microorganisms. Nowadays, predictive models have become a necessary tool to support decisions concerning food safety and quality because models can provide rapid responses to specific questions. Furthermore, predictive models have been incorporated as helpful elements into the self-control systems such as Hazard Analysis for Critical Control Point (HACCP) programs and food safety risk-based metrics. National and international food safety policies are now based on the development of Quantitative Microbial Risk Assessment studies, which is greatly supported by the application of predictive models. Predictive microbiology is still growing but at the same time is turning into an important tool for improving food safety and quality.

Keywords Predictive microbiology • Predictive modeling • HACCP • Growth and inactivation models • Kinetic parameters • Food Safety Objectives • Quantitative Microbial Risk Assessment

F. Pérez-Rodríguez and A. Valero, *Predictive Microbiology in Foods*,
SpringerBriefs in Food, Health, and Nutrition 5, DOI 10.1007/978-1-4614-5520-2_1,
© Fernando Pérez-Rodríguez and Antonio Valero 2013

1.1 Historical Remarks

The origin of predictive microbiology is often linked to the works by Bigelow (1921), Bigelow and Esty (1920), and Esty and Meyer (1922) in which a log-linear model was proposed to describe bacterial death kinetic by heat. Their model found a wide application in the food industry, and especially in the canning industry. Indeed, nowadays these results are still applied by the food industry to reduce *Clostridium botulinum* in low-acid canned foods. As pointed out by McMeekin and Ross (2002), other areas, such as fermentation microbiology, have also contributed to the development of predictive microbiology (Monod 1949). The term predictive microbiology, which is relatively recent, was coined by Roberts and Jarvis (1983), establishing the conceptual basis of modern predictive microbiology. However, many years earkier, Scott (1937) had already put forward similar ideas in a specific case, stressing the importance of knowing growth rates at different temperatures to be able to predict population changes in beef meat.

During the 1960s and 1970s, several efforts were devoted to apply mathematical models to inactivation of pathogens (e.g., *Clostridium botulinum* and *Staphylococcus aureus*) and growth of spoilage bacteria (Spencer and Baines 1964; Nixon 1971). Nonetheless, the great development of predictive microbiology started during the 1980s when computers and specific software facilitated the development of more complex and precise models. McMeekin et al. (1993a) suggested, as another possible explanation, the marked increase in food-borne diseases during those years together with a major awareness of the limitations of the microbiological methods applied in that time, also in consonance with the ideas proposed by Roberts and Jarvis (1983). Those years led to important advances in modeling, proposing different mathematical functions to describe the relationship between temperature and other environmental factors and kinetic parameters (e.g., growth rate). Ratkowsky et al. (1982) introduced a simple model based upon Bêlehrádek-type models to describe the bacteria growth rate as a function of temperature. Other work was concerned with the Arrhenius model and its variations and the cardinal temperature model of Rosso et al. (1995), which generated a new family of models in which the gamma concept model is included (Zwietering et al. 1996). Although these studies provided insight into the underlying mathematical principles governing the dependency of kinetic parameters on environmental factors, other researchers devoted their work to derive suitable mathematical functions to reflect the growth pattern described by bacteria (i.e., cells vs. time) in food environments. In this research area, several models were derived based the sigmoid shape of the microbial growth curve, such as the logistic model and the modified Gompertz equation introduced by Gibson et al. (1987) and the model of Baranyi and Roberts (1994) developed on the basis of the logistic equation, which has come to be one of the most used models together with the reparameterized Gompertz equation (Zwietering et al. 1990). In this productive period of predictive microbiology, some studies aimed to study the probability of growth and toxin production (Roberts et al. 1981; Genigeorgis 1981), which later gave rise to a new type of model, namely, probability or growth/

no growth models for predicting the likelihood that organisms will grow and produce toxin within a given period of time. Only over the past few years has this type of model been more extensively studied to define the absolute limits for growth of microorganisms in specified food environments (Salter et al. 2000; McMeekin et al. 2002; Valero et al. 2010).

During the past few years, predictive microbiology has moved into new research areas such as the study of the effect of pre-culture conditions on kinetic parameters, modeling based on individual cells (Dupont and Augustin 2009), stochastic models, bacterial transfer models (Pérez-Rodríguez et al. 2008), and more recently genome-scale modeling (Brul et al. 2008; Métris et al. 2011). These incipient research areas clearly reflect the continued effort of microbiologists to explore much more deeply, even to genomic level, the ultimate factors or elements governing microbial behavior. However, we should not forget that the final aim of predictive microbiology is to be applied to improving food safety and quality. Thus, the past decade has also seen the great development of Quantitative Microbial Risk Assessment (QMRA) as a fundamental tool to support making decision processes for food safety management underpinned by the use of predictive models (Lammerding and Paoli 1997; Mataragas et al. 2010).

1.2 Framework and Concepts

1.2.1 Predictive Microbiology: Models and Types

Predictive microbiology in foods is a broad scientific field including different concepts and applications. The language used by researchers in this area is sometimes specific and difficult to follow by readers who are not familiar with this terminology. In this section, the predictive microbiology framework is brought to new predictive microbiology practitioners and researchers.

Predictive microbiology can be considered as a scientific branch of the food microbiology field intended to quantitatively assess the microbial behavior in food environments to derive adequate mathematical models. A mathematical model is a description of a real system by using mathematical equations, which are simplifications of the system based on its more significant properties.

A basic model is structured as

The mathematical model estimates the response of the represented system or process based on the values of the input variables. The following generic mathematical function can be used to explain the basic structure:

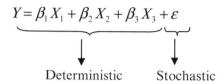

$$Y = \beta_1 X_1 + \beta_2 X_2 + \beta_3 X_3 + \varepsilon$$

Deterministic   Stochastic

The variable Y is the response or dependent variable; X_1, X_2, and X_3 are explanatory or independent variables; and β_1, β_2, and β_3 are the regression coefficients, which are obtained from a regression method based on the observed data set. The error is a term explaining the observed data variability, which is assumed to be normally distributed. Therefore, mathematical models are composed of two components, a deterministic part describing the deterministic relationship between explanatory variables and the response variable and a stochastic part corresponding to the observed data variability that cannot be explained by the deterministic part. The error or stochastic assumption determines the regression method to be used, although in some cases mathematical transformations of variables may help to improve results from a specific regression method. This aspect is further discussed in Chap. 3.

The regression methods are a set of statistical and mathematical techniques intended to fit a mathematical expression to observed data by optimizing the values of the regression coefficients. Optimization of coefficients is based on minimizing the differences between observed responses and the predicted responses given by the fitted model, which is named residual. Several regression methods have been used in predictive microbiology, although the least squares (LS) method is the regression method most used because of its widespread implementation in modeling software. The regression methods largely depend on whether the function to fit shows a nonlinear or linear form. The definition of nonlinearity relates to the regression coefficients and not to the graphical relationship between the variables and the response. For example, the quadratic regression model is considered to be linear rather than nonlinear because the regression function is linear in the regression coefficient and the model can be estimated by using classical linear regression methods.

$$Y = \beta_0 + \beta_1 X + \beta_2 X^2 + \varepsilon \tag{1.1}$$

One of the most common nonlinear models is the exponential decay or exponential growth model, which presents the regression coefficient in the power. In these cases, nonlinear regression methods should be applied to correctly estimate the regression parameters.

$$Y = e^{\alpha X} + \alpha_0 \tag{1.2}$$

Most models developed in predictive microbiology require the application of the nonlinear regression method such as the growth model by Baranyi and Roberts

(1994) or the modified Gompertz equation (Gibson et al. 1987) that describes the bacterial concentration change over time.

Predictive microbial models may describe different microbial processes, including kinetic processes (McMeekin et al. 1993b) such as microbial death and growth, or physical processes such as bacteria and virus transfer (Pérez-Rodríguez et al. 2008), which are described in Chap. 4, although most of the models developed in the scientific literature correspond to kinetic models because of the great repercussions in ensuring food safety and quality.

Traditionally, models have been classified into two types depending on the basis of information used to construct the model (McMeekin and Ross 2002). The mechanistic models are based on understanding the underlying phenomena governing the system. In turn, empirical models simply try to describe the observed response. Besides that, predictive microbiology models describing kinetic process are classified as primary, secondary, and tertiary models. The primary models account for the concentration change versus time (growth or death curve), whereas secondary models relate the kinetic parameters derived from the primary model to environmental variables. The tertiary models are implementations of both types of model in computer tools intended to provide predictions such as the Pathogen Modeling Program (PMP) or Combase Predictor, which are presented in Chap. 5.

In all cases, models should be validated because reliability of predictions should be contrasted with real data in foods. With this end, different strategies have been proposed such as traditional goodness-of-fit indexes applied to observed and predicted data, and more specific indexes such as bias factor and accuracy factors that can be determined by the prediction capacity of models (Ratkowsky 2004). All these concepts and methodology are explained in detail in Chap. 3.

1.2.2 Predictive Microbiology: General Framework

Predictive models are mostly based on observations obtained from experiments developed under controlled conditions. Because experiments with foods are complex and laborious, data are often obtained in simplified experimental models in which the paramount factors are under control. This practice is in accordance with the reductionist approach implicit to predictive microbiology, which claims that microbial behavior can be explained by considering only the most important factors governing the phenomenon (McMeekin and Ross 2002), including not only environmental factors but rather biological factors such as competitive microflora, whose models are so-called between-species interaction models, or natural antimicrobial compounds (Vermeiren et al. 2006; Larsen et al. 2012). Furthermore, over the past few years, there has been an increasing trend to develop predictive models based on data obtained in food with the aim of obtaining more accurate predictions. At the same time, an 'omic' perspective is being introduced into predictive microbiology, moving models onto a molecular level (Brul et al. 2008). These new advances in predictive microbiology are treated in Chap. 7.

Predictive models are useful tools for improving food safety and quality, which can be applied to different facets of the food sector. Among its applications, we highlight its integration into self-control systems such as Hazard Analysis for Critical Control Point (HACCP) to determine process criteria and control limits (McDonald and Sun 1999). Furthermore, applying predictive models can assist shelf life studies and help to design or reformulate food products based on a safety or quality perspective (van Boekel 2008).

The field of Microbial Risk Assessment (MRA) is one of the most relevant topics that has emerged over the decades concerning food microbial safety. MRA is defined as *'a scientifically based process consisting of the following steps: (1) hazard identification, (2) hazard characterization, (3) exposure assessment, and (4) risk characterization'* (Codex Alimentarius Comission 1999). This methodology is the basis for supporting risk management activities and establishing food standards in a national and international framework. A quantitative approach is preferred for MRA, so-called Quantitative Microbial Risk Assessment (QMRA), in which numerical data or quantitative information is used to carry out the foregoing four steps. Briefly, a QMRA consists of quantitatively assessing the fate of a specific pathogenic microorganism along the food chain from farm to fork, and estimating the attendant risk. The development of QMRA studies is greatly supported by the application of predictive models because in many cases no data are available for describing some specific food processes or food chain steps (Lammerding and Fazil 2000). Chapter 6 provides a more specific explanation on the importance of predictive models in QMRA.

The importance of predictive microbiology for supporting the decision-making process concerning food safety can be evidenced by the risk management framework proposed by the International Commission of Microbial Specifications in foods (ICMSF) (van Schothorst 2004). This framework is mainly based on a quantitative approach in which food processes can be defined as increments or decreases of the target microorganism burden. For that purpose, predictive models are the necessary element, which is applied to describe potential microbial increments and decrements along the food chain. By using this quantitative approach and on the basis of established Food Safety Objectives, other risk metrics can be derived such as Performance Objective or Performance Criteria using the ICMSF equation, which will be analyzed in Chap. 6.

1.3   A Tool for Improving Food Safety and Quality

Despite the improvement in food technology and processing, currently microbiological hazards are associated with some food commodities; thus, their evaluation, control, reduction, and/or elimination are important for different commissions, governments, and organisms related to public health.

As previously stated, predictive microbiology is based on the use of mathematical models to estimate microbial behaviour in foods. For food industries, application of this knowledge could be of great interest for assuring food safety and quality.

During the past several years there has been substantial advance in both the concepts and methods used in predictive microbiology. Coupled with 'user-friendly' software and the development of expert systems, these models are providing powerful new tools for rapidly estimating the effects of formulation and storage factors on the microbiological relations in foods.

Thus, predictive microbiology is understood as a scientific-based tool covering an integrated approach that improves food safety and quality. Predictive models can be effectively applied throughout the whole food chain, from raw material acquisition to end products. The utility of predictive microbiology will be further enhanced when is recognized as an effective rapid method (McMeekin et al. 2002).

Bridging the gap between scientific development and practical implementation in industry has always been a challenge. Traditionally, food operators are confident, based on limited information, about their process or product. Indeed, scarcity in available resources and facilities to develop rapid and cost-effective techniques together with the increasing demand of safer and more stable food products from consumers have decreased the investment in research and development sources.

Despite the earlier development of predictive microbiology in the science field, in the 1980s it was accorded more awareness because the outbreaks occurring with traditional food-borne pathogens, such as *Salmonella* spp. in eggs or *Listeria monocytogenes*, which can grow at refrigeration temperatures.

From that time on, national governments and food authorities have prioritized the use of food research for improving food safety.

Some of the potential applications of predictive microbiology are summarized below.

Hazard Analysis and Critical Control Points (HACCP)

- Preliminary hazard analysis
- Identification and establishment of critical control points
- Corrective measures
- Evaluation of variables interaction

Risk Assessment and Risk Management

- Estimation of microbial population dynamics along the food chain
- Exposure assessment toward a specific pathogen
- Design of scientifically based management strategies to assure food safety

Shelf life studies

- Growth prediction of spoilage or pathogenic microorganisms in foods

Innovation and development of a new product

- Evaluation of the impact of microbial spoilage in a food product
- Effect of processing on food quality and safety

- Evaluation of the effect of other additional factors throughout the food chain

Hygienic measures and temperature integration

- Evaluation of the consequences of chill chain application on microbial spoilage
- Optimization of thermal and nonthermal inactivation processes

Education

- Education of both scientific and nonscientific staff
- Implementation and training of computing-based decision systems

Experimental design

- Estimation of the number of samples to be prepared
- Definition of intervals within each factor to be analyzed

Other current applications of predictive microbiology in an industrial context are wide but can be summarized into three groups (Membré and Lambert 2008):

- *Development of new products*: Developing alternative product formulations by the assessment basing on the assessment of growth of spoilage and pathogenic microorganisms; this provides the definition of safer storage conditions, thus increasing shelf life
- *Operational support*: Supporting food safety decisions that need to be made when implementing or running a food manufacturing operation; also, setting critical control points (CCPs) in HACCP, assessing impact of process deviations on microbiological safety and quality of food products
- *Incident support*: Estimating the impact on consumer safety or product quality in case of problems with products on the market

For food industries, besides economical issues, the desire to produce safe foods using strategies is based on understanding sources and magnitudes of hazards.

These are the main reasons why HACCP and risk assessment are taking part in any decision-making process (McMeekin and Ross 2002). Among their main principles, they must assess human exposure to pathogens in foods. Clearly, this information is rarely available. Predictive microbiology models are commonly used to quantify the human exposure of bacteria through the ingestion of foods. To translate these concepts into food safety levels, risk-based metrics are used as a systematic approach to food safety based on hazard identification and control. These metrics (i.e., Performance Objectives and Food Safety Objectives) identify and evaluate key steps in the food production chain, which have the greatest effect on risk associated with hazards; it is often applied subjectively (Stringer 2005).

Thus, predictive microbiology gives improved, quantitative insight into the food properties that are considered of importance to the safety and quality of foods (McMeekin et al. 1993a; Zwietering et al. 1992). Specifically, published studies of potential applications in meat (McDonald and Sun 1999; Sumner and Krist 2002;

Pin et al. 2011) and fish industries (Koutsoumanis and Nychas 2000; Dalgaard et al. 2002; Ross et al. 2003) are described.

Determination of shelf life is one of the most promising applications of predictive microbiology to food industries, rendering a reliable and economic tool for obtaining rapid estimations (Shimoni and Labuza 2000; Castillejo-Rodríguez et al. 2002; Mataragas et al. 2006); this is also related to the applicability of time-temperature indicators for monitoring of spoilage (Vaikousi et al. 2009). In this context, the development and application of structured quality and safety assurance systems based on prevention through monitoring, recording, and controlling of critical parameters during the entire life cycle of the products, seem to be a prerequisite.

More recent applications are focused on the use of predictive modeling to reduce the impact of climate change and seasonal variations on the safety and quality of foods (Janevska et al. 2010). This goal can be reached starting with supply chain data to recalculate periodically changes in model parameters used for prediction of risk levels or shelf life, for example, the probability for contamination of the product with certain pathogens, growth rate, or initial count of spoilage micro-organisms. Application of appropriate statistical analysis would identify significant variations in the trends in terms of decreased safety or shelf life of the product, which would require further attention and corrective actions.

The development of a predictive model comprises different stages, depending on the model type, application, or time/resources to be committed.

Some relevant questions are often arisen before the construction of a new or modified predictive model. As literature sources contain a large quantity of developed models, much information is already gained by what one has really to analyze if carrying out additional experimental trials to create new models regarding a particular hazard/food combination.

Despite each particular situation, general questions (Q) could be formulated:

Q1: Are data and/or models sufficiently available in the literature?
Q2: Will the predictive model significantly improve current knowledge in the field?
Q3: Are available laboratory resources sufficient to perform all analyses in controlled conditions?
Q4: Do the authors have 'a priori' knowledge about the main environmental conditions affecting microbial growth/survival of the studied hazard(s) and food(s)?
Q5: According to these conditions, is it possible to develop a full-factorial design?
Q6: Is the mathematical model comprehensive and representative of the observed behaviour of the studied hazard(s)?
Q7: Could the model be validated with additional measurements or external data?
Q8: Could the model be effectively applied for food industries or authorities under a given set of conditions?

In the decision tree presented in Table 1.1, according to yes (Y) and no (N) answers, one can decide if a model can be applied.

Throughout this brief, each of these steps is analyzed accordingly. However, experimental design is one of the most important because it will influence the type

Table 1.1 Aspects to be considered when constructing and applying a predictive model

STEP 1		
Literature review and preliminary analysis		
Q1	Y	Discard
	N	Next step
Q2	Y	Next step
	N	Depends on the model purpose
STEP 2		
Planning the experimental design		
Q3	Y	Next step
	N	Redefine the experimental design
Q4	Y	Next step
	N	Go to STEP 1
Q5	Y	Next step
	N	Create a non-full-factorial experimental design
STEP 3		
Model development		
Q6	Y	Next step
	N	Revise data processing and choose another equation
STEP 4		
Model validation		
Q7	Y	Next step
	N	Go to STEPS 1 and 2
STEP 5		
Model application		
Q8	Y	Application
	N	Go to STEPS 3 and 4

and number of conditions tested, data gathering, data processing, and definitely the final model obtained. Without an adequate experimental design, data generated will be erratic, thus needing further repetitions. The next chapters consider experimental analyses and procedures for data generation.

Chapter 2
Experimental Design and Data Generation

Abstract One of the most critical steps when generating a predictive model is to correctly design an experiment and collect suitable microbial data. Experimental design will influence model structure and validation conditions. The survival and growth of microorganisms in foods is affected not only by the chemical composition of the food and its storage conditions but also by the food matrix. In this sense, a better quantification of the food structure effect has been studied throughout these years. Regarding the method of data collection, although plating count has been widely used (and still is used), there are rapid methods to obtain reliable and cost-effective data. These achievements were primarily based on turbidimetry, although other methods (microscopy, image analysis, flow cytometry, etc.) have arisen as novel approaches in the predictive microbiology field. These aspects are further discussed in this chapter.

Keywords Experimental design • Food matrix • Challenge testing • Data generation • Absorbance • Turbidimetry • Flow cytometry • Microscopic methods

2.1 Experimental Design

In predictive microbiology, as with other scientific disciplines, collection of high-quality data is an essential part of exploitation of results. Both the selection of an appropriate model structure and the identification of accurate model parameters are data-driven processes; that is, the efficiency and accuracy of these procedures are determined by the quality of the experimental data.

The experimental design of a predictive model will mainly depend on its final application into a real case scenario. This process is completely different when estimating growth kinetic parameters as a function of certain environmental factors than when one qualitatively estimate the probability that a given microorganism may or may not grow under a specific set of conditions. Similarly, when performing a validation study in a food matrix, design will be accomplished according to the

F. Pérez-Rodríguez and A. Valero, *Predictive Microbiology in Foods*,
SpringerBriefs in Food, Health, and Nutrition 5, DOI 10.1007/978-1-4614-5520-2_2,
© Fernando Pérez-Rodríguez and Antonio Valero 2013

representativeness of data to real conditions or to the time invested in analytical experimentation. For this purpose, it is very useful to previously screen the main factors affecting microbial behavior through different assays. Implicitly, when increasing the number and levels of involved factors, the experimental design will be more complex.

Devlieghere (2000) already described the main factors to be considered to plan an adequate experimental design. Some relevant questions are these:

- What is the main objective of the predictive model?
- Which are the main factors to be controlled, so that this objective will be achieved?
- Which are the factor levels to be used?
- Which is the inoculum state to be employed? This refers to physiological state, use of cocktail strains, inoculation form, etc.
- Which are the dependent variables of the proposed model?
- Which is the substrate or medium used?
- How many combinations of environmental factors will be finally included in the model (from those previously identified)?
- How will data be collected?

Some of this information is discussed in the following sections. In the meanwhile, referring to the experimental design, a two-step procedure is often applied: (1) screening experiments are performed at an extended range of the factors, followed by (2) an extensive data collection study within the region of interest (Gysemans et al. 2007). This latter point is referred to the inclusion of additional levels of identified factors to obtain a more refined model and consequently, more accurate microbial predictions.

In modeling microbial responses in foods, traditionally, full experimental designs are chosen (Tassou et al. 2009). This approach considers all combinations of the different explanatory variables. The main advantages are its ease of implementation together with its data processing. Further, one is sure that all information is gained from the experiment, as all combinations are explored. However, this experimental design is often labor intensive and costly.

Alternatives are being adopted by carefully selecting experimental conditions (implementing an efficient design-of-experiment, or DOE). The number of experiments required for it is calculated as $(N) \sim 2^k$, where (k) is the number of variables.

If the number of variables is large, the fractional factorial design can be more indicative.

Fractional factorial designs are reduced versions of full factorial designs, and some of these have been published (McKellar and Lu 2001; Valero et al. 2009). These designs are based on 'a priori' knowledge or assumptions on the most important factors or expected interactions. A particular class of fractional factorial designs called the Box–Benhken designs has been used for modeling microbial growth or inactivation. Combining two-level factorial designs with balanced incomplete blocks forms this experimental design. The repeatability of the model

Fig. 2.1 Schematic
representation of a four-factor
level Latin-Square design

A	B	C	D
C	D	A	B
D	C	B	A
B	A	D	C

Fig. 2.2 Schematic
representation of a two-factor
Central Composite design

is normally tested through the performance of additional experiments at the central points of the experimental design.

Latin-Square designs are special types of fractional factorial designs. According to its definition, a Latin square of order x is an arrangement of x letters in an x-by-x array so that each letter appears exactly once in each row and exactly once in each column. In the context of experiment design in predictive microbiology, one primary (treatment) factor (represented by the letters) is typically studied in the presence of several blocking (nuisance) factors, although the approach is not limited to one principal factor of interest. Latin-Square designs can be extended to more individual factors, such as the Graeco-Latin Square or the hyper Graeco-Latin Square designs. A Latin-Square design of order 4 is presented in Fig. 2.1.

Another experimental design commonly used in the field of predictive microbiology, the Central Composite Design (Cheroutre-Vialette and Lebert 2002; Arroyo-López et al. 2009), contains an embedded factorial or fractional factorial design with center points that is augmented with a group of 'star points' which allow estimation of curvature. If the distance from the center of the design space to a factorial point is ± 1 unit for each factor, the distance from the center of the design space to a star point is $\pm\alpha$ with $\| \alpha > 1$. The precise value of α depends on certain properties desired for the design and on the number of factors involved. If two factors are included, the Central Composite design will appear as shown in Fig. 2.2.

The Doehlert matrix describes a spherical experimental domain and stresses uniformity in space filling. For two variables it consists of one central point and six points forming a hexagon, situated on a circle. The formula used for calculation of the number (N) of experiments required is ($N \sim k^2 + k + C_0$), where (k) is the number of variables and (C_0) is the number of center points. Replicates at the central level of the variables are performed to validate the model by means of an estimate of experimental variance.

For rather simple model structures and a limited number of levels per environmental factor, full factorial designs are preferable because these designs guarantee accurate and reliable model parameters (Mertens et al. 2012). However, for more complex cases, a Latin-Square design can be considered as an attractive alternative as it does not require a priori knowledge of the model structure (as is the case for the reduced full factorial design), while keeping the experimental workload and cost low. In contrast, central composite designs should be avoided because of the high degree of uncertainty on the parameter estimates.

2.2 Growth Matrix: Food Versus Artificial Medium

For a long time, the ability of specific foods to support microbial growth of pathogenic as well as spoilage microorganisms has been evaluated by inoculating the target organism and monitoring its growth and survival over a certain period of time. This methodology, traditionally named challenge testing, is still being used in the field of predictive microbiology because it is sometimes necessary to gain information about the microbial stability of a novel product formulation or to assess the behavior of a specific microorganism (not previously tested). These experimental tests are useful to determine microbial shelf life and growth/survival kinetics parameters, such as maximum growth rate or lag phase. However, although this approach was considered the gold standard, it is also time consuming and costly. Thanks to the development of predictive microbiology, microbial behavior is explained with only a few significant environmental factors (mainly temperature, pH, or a_w), thus yielding accurate predictions in the majority of foods.

Most published studies of predictive microbiology (at least the earlier ones) used artificial media, that is, culture media with a chemical composition that is intended to mimic the food environment, which allowed a reduced variability in the results (mainly because chemical composition can be more accurately adjusted). Also, it is recognized that artificial liquid media provides a more homogeneous distribution of microorganisms, leading to obtaining similar kinetics under the same environmental conditions.

In principle, observed data can be easily fitted to mathematical models because they tend to be more robust and replications do not vary greatly.

A key step before the application of predictive models is validation in food matrices: this involves the comparison of model predictions with additional data coming from literature sources or by means of inoculating the target organism in a given food (supposedly within the range of conditions covered by the model) and evaluating the observations with model predictions to judge if they are biased. Throughout this brief, a specific section about model validation is proposed, with special attention to validation procedures, indexes used, and reliability of models when applied to food matrices.

In this section, differences between structural composition between artificial media and foods are presented.

Traditionally, when comparing model predictions in broth media with observations in foods, one can assume that results will be fail-safe, that is, the predicted growth in liquid media is much faster than that observed in food. Several factors are attributable: food matrix (which in most cases is semisolid or solid),

Fig. 2.3 Representation of the most common spatial distributions of microbial populations according to the type of matrix in which they are present

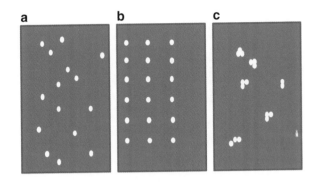

indigenous flora, or additional environmental factors present in the food and not included in the model.

Growth of microorganisms in a liquid aqueous phase in foods is typically planktonic and can be accompanied by motility allowing taxis to preferred regions of the food. Transport of nutrients to the bacteria and of metabolites away from them results in a locally uniform environment until considerable accumulation of microbial biomass and metabolites cause bulk chemical changes (such as a decline in pH). It is this microarchitecture in food that is mimicked in microbiological experiments by the use of broth culture medium (Wilson et al. 2002).

However, foods are not typically homogeneous. The structure of foods creates local physical or chemical environments that clearly affect spatial distribution of microorganisms. Consequently, microbial growth or survival is influenced. Microorganisms occupy the aqueous phase of foods, and structural features of this phase cause an effect on microbial behavior such as constraints of mechanical distribution of water, redistribution of organic acids, and food preservatives (Hills et al. 1997).

Predictions based on data obtained from broth systems can be successfully applied to microbial growth and survival in foods. In many cases, microorganisms grow more slowly in a given structure food than in broths: this is the case when performing a challenge testing, in which lower growth rates (and longer lag phases) are observed than are proposed by predictive models (Pin et al. 1999). In structured (heterogeneous) foods, microbial growth can strongly depend on the position in the food and the assumption of perfect mixing can thus not be accepted. In consequence, space must be considered (Dens and Van Impe 2000).

This concept is related to the microbial distribution of microorganisms. Figure 2.3 represents the distribution of a microbial population in a specific matrix (International Life Sciences Institute 2010). Normally, in liquid media random distribution is obtained, indicating that there is equal chance for any individual cell to occupy any specific position. Presence of organisms can be considered independent (Fig. 2.3a). Regular distribution is observed when high cell densities are encountered in the food and these cells are neither clumped nor aggregated (Fig. 2.3b), as when contamination occurs in equipment or utensils when they are insufficiently cleaned. Finally, in solid matrices colonies are normally are aggregated, forming clusters (biofilms): the presence of organisms can be considered dependent.

Once a food product becomes contaminated, microbial growth can transform the initial homogeneous distribution into a more clustered one. When cells are growing inside the product, the spatial distribution is more similar to that presented in Fig. 2.3c, mainly as a result of the physical constraints of the food matrix.

Regarding inactivation treatments, it is demonstrated that application of heat may have a different impact on microbial death depending on the considered food matrix. If thermal diffusivity of the product is lower, some contaminants can survive if the heat has not sufficiently conducted to the interior of the food product. This formation of the so-called cold spots may affect the spatial location of microorganisms.

Thus, the use of model foods that mimic real food structures entails significant advantages from a practical point of view, such as better control, ease of operation, and repeatability of analyses (Antwi et al. 2007; Noriega et al. 2008).

2.3 Data Generation

Data generation must be based on the optimization of the number of data together with the implied cost of their acquisition. Generally, when building a growth and survival model it is known that data should be collected throughout the whole analytical period of time, about 100 kinetic curves being needed to make the model significant (McDonald and Sun 1999). Gibson et al. (1987) concluded that 15 points per kinetic curve are necessary, 20 being the optimum number. Less than 10 points per curve makes the adjustment not fully representative of microbial behavior, thus increasing uncertainty. Poschet and Van Impe (1999) established that this uncertainty on dependent variables increases when fewer points are collected, but, above a certain limit, this value remains stable.

Similarly, distribution of collected points within the experimental design is crucial for estimation of dependent variables: this is achieved in such a way that the representativeness of the model increases, at the same time reducing variance of the estimated parameters.

There are currently different methods to collect data, but the classical one has been widely used in predictive microbiology, that is, plate count techniques. This method has been used to monitor microbial growth although it presents some drawbacks. First, possible underestimation could occur because of the presence of clumped cells. Second, plate count is a slow, labor-intensive, and costly method. Further, it does not provide immediate results, so that model development becomes more difficult, especially when a high number of data is required. However, to a certain extent, plate counts can be done automatically by the use of automated platers such as the spiral plater and automatic colony readers.

Traditionally, models performed in broth systems are based on the modification of artificial culture media with preservatives, such as organic acids, sodium chloride, or sodium nitrite, or on lowering pH with the addition of chlorhydric acid. Afterward, these modified media are inoculated with the specific microorganism, and growth and survival are monitored over a certain period of time.

When concerned with food matrices, one way to assess microbial behavior is to perform a challenge test, that is, to inoculate the target microorganism in a specific food and to evaluate if the food supports growth and survival under a limited set of conditions.

The next section explains in more detail the experimental procedure to carry out a general challenge test. Of course, there are many variants and uses of this method, in accordance with the final objective to be fulfilled. This explanation summarizes the most generic steps to achieve a challenge test.

2.3.1 Traditional Methods: Challenge Testing

Microbiological challenge testing has been and continues to be a useful tool for determining the ability of a food to support the growth of spoilage organisms or pathogens (US Food and Drug Administration 2009). In predictive microbiology, it is needed to evaluate the behavior of a particular strain (or a cocktail of different strains) to subsequently calculate kinetic parameters. Microbiological challenge tests also play an important role in the validation of processes that are intended to deliver some degree of lethality against a target organism or group of target organisms (for example, a 5 log reduction of *Escherichia coli* O157:H7 for fermented meats). Selection of microorganisms to use in challenge testing and/or modeling depends on the knowledge gained through commercial experience and/or on epidemiological data that indicate that the food under consideration or similar foods may be hazardous because of pathogen growth.

An appropriately designed microbiological challenge test will validate that a specific process is in compliance with the predetermined performance standard. The design, implementation, and assessment of microbiological challenge studies form a complex task that depends on factors related to how the product is formulated, manufactured, packaged, distributed, prepared, and consumed.

Failure to account for specific product and environmental factors in the design of the test could result in flawed conclusions.

Microbiological challenge studies can be used in specific cases for the determination of the potential shelf life of certain refrigerated or ambient-stored foods. The determination of whether challenge studies are appropriate or useful must consider such factors as the likelihood of the product to support growth of spoilage organisms or pathogens or include knowledge of the previous history of the product.

When conducting a microbiological challenge study, a number of factors must be considered: (a) the selection of appropriate pathogens or surrogates, (b) the inoculation level, (c) the method of inoculation, (d) the duration of the study, (e) formulation factors and storage conditions, and (f) sample analyses. These are described next.

(a) Selection of appropriate strains or surrogates
    The ideal organisms for challenge testing are those that have been previously isolated from similar formulations. Additionally, pathogens from known foodborne outbreaks should be included to ensure the formulation is robust enough to inhibit those organisms as well.

For certain applications, surrogate microorganisms may be used in challenge studies in place of specific pathogens. For example, introducing pathogens into a processing facility is not feasible; therefore, it is desirable to use surrogate microorganisms in those cases. An ideal surrogate is a strain of the target pathogen that retains all other characteristics except its virulence.

In any case, it is important to incubate the microorganisms in standardized conditions, preferably similar to those encountered in the food.

(b) Inoculation level

The inoculum level used in the microbiological challenge study depends on whether the objective of the study is to determine product stability and shelf life or to validate a step in the process designed to reduce microbial numbers. Typically, an inoculum level between 10^2 and 10^3 colony-forming units (cfu)/g of product is used to ascertain the microbiological stability of a formulation. If the inoculum level is too low, microorganisms could not grow in the food product because of the increased lag phase, so that one can assume in certain cases that food formulation assures food safety when it is not low. In contrast, at high inoculum levels, microbial growth could be overestimated. For studying lethality processes, higher levels of microorganisms are needed (generally more than 10^6 cfu/g).

(c) Method of inoculation

The method of inoculation is another extremely important consideration when conducting a microbiological challenge study. Every effort must be made not to change the critical parameters of the product formulation undergoing challenge. A variety of inoculation methods can be used depending upon the type of product being challenged. In aqueous liquid matrices such as sauces and gravies with high a_w (>0.96), the challenge inoculum may be directly inoculated into the product with mixing, using a minimal amount of sterile water or buffer as a carrier. Use of a diluent adjusted to the approximate a_w of the product using the humectant present in the food minimizes the potential for erroneous results in intermediate a_w foods. In studies where moisture level is one of the experimental variables, the inoculum may be suspended in the water or liquid used to adjust the moisture level of the formulation. Products or components with $a_w < 0.92$ may be inoculated using the atomizer method with a minimal volume of carrier water or buffer. Again, the product should always be checked to ensure that the final product a_w or moisture level has not been changed. A short postinoculation drying period may be needed for some products before final packaging. A minimum volume of sample should be inoculated so that a minimum of three replicates per sampling time is available throughout the challenge study. In some cases, such as in certain revalidation studies and for uninoculated control samples, fewer replicates may be used.

(d) Duration of the study

The microbiological challenge test should be conducted for at least the whole shelf life period of the food product. Some regulatory agencies recommend extending the duration of the study a margin beyond the desired shelf life because it is important to determine what would happen if users held the product beyond its intended shelf life and then consumed it.

The frequency of analysis depends on the environmental conditions under which the food is subjected. It may be desirable to test more frequently (for example, daily or multiple times per day) early in the challenge study (that is, for the first few days or week) and then reduce the frequency of testing to longer intervals.

(e) Formulation factors and storage conditions

When evaluating a formulation, it is important to understand the range of key factors that control its microbiological stability. It is, therefore, important to test each key variable singly or in combination in the formulation under worst-case conditions. Experimental temperature should be similar to real processing, distribution, and sale conditions. In a last step, the use of temperature shift might be recommended, such as storing the food product at one specific temperature for a portion of its shelf life, after which time the product may be subjected to elevated temperatures.

(f) Sample analysis

In challenge tests, it is recommended to analyze at least three replicates per analytical point, although more replicates would be needed when requiring more accuracy in the results. The culture media to be used will depend on the type of microorganism to be controlled, but if the food product contains high concentration levels of competitive flora, it would be better to use selective media. Similarly, if the targeted microorganism is a toxin producer, toxin concentration should be measured during the study period. In parallel, control samples (uninoculated) may be analyzed in the same way as inoculated samples to evaluate the effect of the food flora on the analytical period of time.

2.3.2   *Rapid Methods*

(a) Viable counts

Viable counts are commonly obtained by spread-plate and pour-plate techniques and therefore are linked to the classical microbiological methods that are considered to be reference methods, even though these can have certain limitations (Rasch 2004). For instance, a clump of cells falling onto culture agar would give rise a single colony, impeding estimation of the actual number of individual cells when clumps are present in the medium. Hence, microbiologists refer to results as colony-forming units (cfu) (McMeekin et al. 1993a). Although automation has been also applied to plate count methods (e.g., the spiral plating method) reducing sample preparation time (e.g., decimal dilutions, human resources, etc.), this methodology still implies long waiting times for enabling bacterial growth in culture media (24–48 h). Predictive models can be developed based on data obtained in liquid and solid media or on food matrices. When models are developed on food, more additional steps such as a homogenization step or filtering are required, making the analytical process more laborious. Also, the medium where data are being generated can affect the precision and accuracy of plate counts, which should be considered when predictive models are built, although it is generally accepted that the repeatability of enumeration data may only be precise to about ±0.5 log cfu (Mossel 1995).

(b) Turbidimetry

Bacterial kinetic modeling is mainly based on colony counting (traditional method) and absorbance measures. It is well known that cellular concentration in a liquid medium can be related to the optical density (OD) of the growth medium. OD, or absorbance, is a measure of the light that is absorbed by a cellular suspension. The chief characteristic of OD is that OD of a bacterial suspension increases proportionally with the increment of bacterial concentration. OD of a cellular suspension can be also related to transmittance and turbidity of the medium, important optical properties related to bacterial concentration. Other factors affect OD such as the refractive index of the bacterial strain and its shape and size. The main advantages of OD are rapidity, simplicity, and noninvasiveness, which make the technique quite suitable for modeling purposes. Automation in measuring OD has led to sophisticated photometers such as Bioscreen C, which are able to perform multiple measurements (200 wells) at specific time intervals while maintaining a fixed incubation temperature. Nevertheless, important limitations exist that should be considered when OD is used as the enumeration technique for modeling bacterial growth. Linearity may be one of the most important drawbacks because the linear relationship described by Beer-Lambert only holds for approximately a tenfold increase in cell numbers:

$$\log\left(\frac{I_{\text{incident}}}{I_{\text{transmitted}}}\right) = \text{absorbance} = -x\, c\, l \tag{2.1}$$

here I_{incident} is the intensity of light entering the medium and $I_{\text{transmitted}}$ is the intensity of light exiting the medium, x is a constant dependent on the medium and microorganism, c is microorganism concentration, and l is the distance the light travels through in the medium (i.e., light path).

In some cases, nonlinearity of OD response has been corrected using an empirical function derived from specific experiments in which bacterial suspensions with concentration levels outside the range of linearity are diluted and absorbance is measured. The relationship found between initial absorbance and absorbance after dilution is analyzed to derive a correction function for nonlinear OD data (Dalgaard et al. 1994).

Another important drawback associated with measuring OD is its relatively high detection limit, which is often above 6 log cfu/ml, meaning that a growth model should be based on high initial inoculum levels. The impossibility of differentiating between living and dead cells limits its application to growth models.

In spite of all these limitations, predictive models based on OD data are very often reported in the scientific literature, probably because fewer experimental resources are needed to assay the multiple environmental conditions needed for building secondary models. Moreover, kinetic parameters such as growth rate can be estimated only based on OD measures because this parameter expresses a rate of change with respect to time. Several studies have successfully modeled these parameters based on OD data for different microorganisms and environmental conditions (Dalgaard et al. 1994; Begot et al. 1996; Augustin et al. 1999). However, some studies have proved that models based on OD data can underestimate maximum growth rate, recommending the use of the detection time approach for better estimation of this parameter (Lindqvist 2006; Lianou and Koutsoumanis 2011). The detection time (DT) approach consists of performing several decimal dilutions of the initial inoculum and estimating the DT based on OD data. Then, DTs are used to fit the following equation, which provides an estimation of maximum growth rate (Cuppers and Smelt 1993; Lianou and Koutsoumanis 2011):

$$\log(N_i) = k - \mu_{max} DT_i \tag{2.2}$$

where N_i is the inoculum size corresponding to different decimal dilutions of the initial inocula, μ_{max} is the maximum growth rate, and k is a constant.

The use of calibration curves relating OD and bacterial concentration also produces reliable estimates of kinetic parameters and is an alternative to using OD directly. Calibration equations estimated from experimental data are used to transform OD values to count data. Then, the estimated counts can be used to estimate kinetic parameters (Dalgaard and Koutsoumanis 2001). Precautions should be taken when environmental stresses are applied during bacterial growth because they can affect OD measures (i.e., bacterial shape and size); therefore, calibration curves should be done for each specific growth condition (Valero et al. 2006).

The development of growth/non-growth models is mainly based on use of the OD technique because some of the limitations of the OD technique are not given for this type of study in which no growth rate is observed but only if growth takes place. That condition is experimentally determined based on recording a significant change of OD in the microorganism suspension that is related to growth (Salter et al. 2000; Skandamis et al. 2007; Valero et al. 2009, 2010).

(c) Flow cytometry
Flow cytometry is a rapid technique based on labeling cells in suspension with fluorochrome molecules and passing them, in a liquid stream, through a microcapillary equipped with an electronic detection apparatus. The characteristics of light scattering, light excitation, and emission of fluorochrome from cells are collected to provide information on physiological state, size, shape, and integrity of the analyzed cells. Moreover, the technique can be used to enumerate target microorganisms, showing a good correlation with colony-counting methods (Sørensen and Jakobsen 1996; Endo et al. 2001; Holm et al. 2004).

The combination of both applications, that is, cell enumeration capacity and cell physiological characterization, makes this technique an excellent method to study and model microbial population heterogeneity, which is particularly relevant under stress conditions (Fernandes et al. 2011). In spite of its promising application in predictive microbiology, few studies have been carried out based on data obtained by flow cytometry (Sørensen and Jakobsen 1996; Ferrer et al. 2009). It is likely that further development of omic models, based on a molecular level approach, and new technological advances in cytometry will boost the application of flow cytometry in predictive microbiology studies in future years.

(d) Microscopy and image analysis

This method offers some advantages when compared to plate counting methods and enumeration methods based on optical density. One of the most important advantages is that direct observation on food matrices or artificial media enables obtaining a lower limit of quantification, improving accuracy and the precision of results. In addition, based on the biochemical properties of cells and the use of specific fluorochromes, this technique can provide information on the physiological state of bacteria on the surface (alive/dead, sterease activity, etc.) (Bredholt et al. 1999). More recently, microscopy has been applied to investigate and model individual cell lag times based on observation of systems containing isolated cells (Métris et al. 2005; Niven et al. 2006; Stringer et al. 2011; Gougouli and Koutsoumanis 2012). In general, these systems consist of a surface inoculated with the test microorganism, which can be agar or a microscope slide placed within a tailor-made chamber or a device enabling control of environmental conditions such as temperature or atmosphere (Métris et al. 2005; Niven et al. 2006). A photographic camera is usually coupled with the microscope to capture images of cellular division at certain intervals of time. Moreover, specific software such as ImageJ (Abramoff et al. 2004) should be applied to analyze images taken by the camera, obtaining counts and estimating lag time.

Image analysis can be also applied to estimate kinetic parameters based on radial growth of colonies of bacteria (Dykes 1999; Guillier et al. 2006). However, this methodology is most preferred to model kinetic parameters (i.e., growth rate and lag time) of molds, monitoring the radial growth of mycelium (Rosso and Robinson 2001; Baert et al. 2007; Garcia et al. 2010). Predictive model studies have involved different fungal species such as *Botrytis cynerea*, *Penicillium expansum*, and *Aspergillus carbonarius* (Dantigny et al. 2007; Judet-Correia et al. 2010). In specific cases, image analysis supported by suitable software has been also used to evaluate mycelium growth (Judet-Correia et al. 2010). Automation of this process via image-analyzing systems would further facilitate the application of this method to generate more reliable predictive models.

(e) Electrochemical methods: impedance and conductance.

This technique is based on the fact that bacteria during growth produce positively and negatively charged chemical compounds that modify the impedance of growth medium (Rasch 2004). The time at which a significant change of impedance in the growth medium is detected is the so-called detection time (DT), which is inversely proportional to the logarithm of the initial concentration level of the microorganism (Jasson et al. 2010). Also, conductance and capacitance of growth medium can be used to enumerate bacteria in culture media and foods (Lanzanova et al. 1993; Noble 1999; Koutsoumanis and Nychas 2000). Besides their application for microorganism enumerations, impedance data can be directly used to derive kinetic parameters (McMeekin et al. 1993a). Some examples of this have been successfully developed for *Salmonella enteritidis*, acid lactic bacteria, and *Yersinia enterocolitica*, in which kinetic parameters have been derived from fitting primary growth models (e.g., Gompertz model) to conductance or impedance data and using DT with a similar mathematical approach to that used in optical density methods (Lanzanova et al. 1993; Lindberg and Borch 1994; Fehlhaber and Krüger, 1998).

Chapter 3
Predictive Models: Foundation, Types, and Development

Abstract According to their structure, predictive models can be primary, secondary, or tertiary. This classification mainly depends on the final purpose and type of prediction generated. There has been a significant evolution in the past few years toward better understanding of microbial behavior in foods. Therefore, models that describe the biological process of microbial growth and inactivation have been subsequently developed. Also, fitting methods for linear and nonlinear regression together with goodness-of-fit indexes give us useful information about how the model is able to explain the observed data. Finally, models cannot be applied if a validation process is not previously accomplished, which typically consists of confirming the predictions experimentally by using any quantitative method. In this chapter, a comprehensive review of the most popular validation methods is provided.

Keywords Mathematical function • Mechanistic models • Dynamic conditions • Regression • Polynomial models • Artificial Neural Networks • Fitting methods • Goodness of fit • Validation

3.1 Introduction

It is a goal of food microbiologists to know in advance the behavior of microorganisms in foods under foreseeable conditions. To do so, an exhaustive control of physicochemical factors that could influence microbial growth is needed (such as temperature, pH, a_w, salt), as well as a deep knowledge about the biological characteristics of the target microorganism(s).

As stated in the preceding chapter, the premises behind the scientific basis of predictive microbiology are that microbial responses in foods are in certain way reproducible against several extrinsic and intrinsic environmental factors (Ross et al. 2000). This behavior can be translated into diverse mathematical models that estimate microbial growth/inactivation/toxin production/probability of growth, etc. This emerging area was redefined recently as modeling of microbial responses

F. Pérez-Rodríguez and A. Valero, *Predictive Microbiology in Foods*,
SpringerBriefs in Food, Health, and Nutrition 5, DOI 10.1007/978-1-4614-5520-2_3,
© Fernando Pérez-Rodríguez and Antonio Valero 2013

in foods. McKellar and Lu (2004) presented a detailed review of the predictive models published so far.

With the aim of performing comparative studies, several authors suggested different classifications of predictive models based on their final purpose, the type of microorganism to be studied, and their impact on food spoilage or food safety. Basically, predictive models are split up into three groups: survival/inactivation models, boundary (growth/no growth) models, and growth models. Basing on their development, models can be classified as follows:

(a) Primary models: aim to describe the kinetics of a process with as few parameters as possible while being able to accurately define the growth and inactivation phases. They are represented as the increase (or decrease) in population density against time.

(b) Secondary models: describe the effect of environmental conditions (i.e., physicochemical and biological factors) on the values of the parameters of a primary model.

(c) Tertiary models: based on computer software programs that provide an interface between the underlying mathematics and the user, allowing model inputs to be entered and estimates to be observed through simplified graphical outputs.

Whiting and Buchanan (1994a) called the foregoing integrated software-based models 'tertiary models.' They defined tertiary-level models as personal computer software packages that use the pertinent information from primary- and secondary-level models to generate desired graphs, predictions, and comparisons. Primary-level models describe the change in microbial numbers over time, and secondary-level models indicate how the features of primary models change with respect to one or more environmental factors such as pH, temperature, and a_w. Following is a description of the most relevant primary and secondary models as well as their uses and scope.

3.2 Primary Models

As already described, primary models intend estimating kinetic parameters (e.g., maximum growth rate, lag phase, inactivation rate) as a function of treatment time. For model application, this time can assume, for instance, a storage phase, processing, and/or thermal treatment.

The application of primary models to a set of microbiological data proceeds as follows. In a first step, a mathematical model is assumed to explain the data, that is, how microbial counts change over time. In a second step, such a model is fitted to microbiological data by means of a regression (linear or nonlinear). As a consequence of the fitting process, a number of kinetic parameters embedded in the model is provided, such as the rate of growth/inactivation or lag time (in growth processes) or 'shoulder'/'tail' (in inactivation processes). The dataset used to fit the model is normally obtained under specific intrinsic and extrinsic factors. For this reason, the kinetic parameters provided after fitting solely apply for the specific intrinsic and extrinsic factors characterizing the dataset.

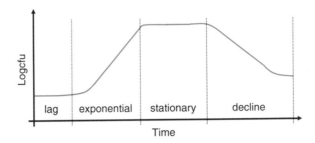

Fig. 3.1 Representation of the four-phase kinetics followed by a generic microbial population in a medium

As represented in Fig. 3.1, primary models aim to describe the four phases of a typical microbial population. Lag phase or the adaptation period is defined as an adjustment period during which bacterial cells modify themselves to take advantage of the new environment and initiate exponential growth (Buchanan and Klawitter 1991). Then, microorganisms grow exponentially (exponential phase) until they reach a 'plateau' or maximum population level (stationary phase). When concentration of nutrients or physiological state of cells is decreasing, the microbial population starts to decline (decline phase). There is here a distinction between 'survival' and 'inactivation' models. In brief, survival is understood as a process or environmental condition not intentionally designed to kill the microbial population but also to not allow growth (e.g., bacteriostatic), whereas inactivation is a process implemented to destroy a microbial population in a specific food by certain log numbers.

Several primary models have been published. Next are described the most important models used for growth and survival/inactivation.

3.2.1 Growth Models

3.2.1.1 Sigmoid Functions

The equations introduced by Gibson et al. (1987), such as the modified logistic (Eq. (3.1)) and the modified Gompertz (Eq. (3.2)) have been those most popular to fit microbial growth data because these functions consist of three phases, similar to the microbial growth curve:

$$\log x(t) = A + \left[C / \left(1 + e^{(-B(t-M))} \right) \right] \tag{3.1}$$

$$\log x(t) = A + C \exp\{-\exp[-B(t - M)]\} \tag{3.2}$$

where $x(t)$ is the number of cells at time t, A is the lower asymptotic value as t decreases to zero, C is the difference between the upper and lower asymptote, M is the time at which the absolute growth rate is maximum, and B is the relative growth rate at M.

The parameters of the modified Gompertz equation (A, C, B, and M) can be used to characterize bacterial growth as follows:

$$lag = M - \frac{1}{B} + \left[\frac{(\log N_0 - A)}{\left(\frac{BC}{e}\right)} \right] \tag{3.3}$$

$$\text{Specific growth rate} = \frac{BC}{e} \gg \frac{BC}{2.18} \tag{3.4}$$

To simplify the fitting process, reparameterized versions of the Gompertz equation have been proposed (Zwietering et al. 1990; Willocx et al. 1993):

$$\log x = A + C \exp\left\{ -\exp\left[2.71 \left(\frac{R_g}{C} \right) (t_{lag} - t) + 1 \right] \right\} \tag{3.5}$$

where $A = \log x_0$ (log cfu/ml), x_0 is the initial cell number, C the asymptotic increase in population density (log cfu/ml), R_g the growth rate (log cfu/h), and t_{lag} is the lag time duration (h).

Although it is used extensively, some authors (Whiting and Cygnarowicz-Provost 1992; Baranyi 1992; Dalgaard et al. 1994; Membré et al. 1999) reported that the Gompertz equation systematically overestimated growth rate compared with the usual definition of the maximum growth rate. Nevertheless, this model was for a long time that most widely used to fit bacterial curves.

3.2.1.2   Mechanistic Functions

Baranyi et al. (1993, 1995) and Baranyi and Roberts (1994) introduced a mechanistic model for bacterial growth (Eq. (3.6)). In this model, it is assumed that during lag phase bacteria need to synthesize an unknown substrate q critical for growth. Once cells have adjusted to the environment, they grow exponentially until limited by restrictions dictated by the growth medium.

$$\frac{dx}{dt} = \frac{q(t)}{q(t) + 1} \cdot \mu_{\max} \cdot \left(1 - \left(\frac{x(t)}{x_{\max}} \right)^m \right) x(t) \tag{3.6}$$

where x is the number of cells at time t, $x_{\max}$ the maximum cell density, and $q(t)$ is the concentration of limiting substrate, which changes with time (Eq. (3.7)). The parameter m characterizes the curvature before the stationary phase. When $m = 1$, the function reduces to a logistic curve, a simplification of the model that is often assumed.

$$\frac{dq}{dt} = \mu_{\max} \cdot q(t) \tag{3.7}$$

The initial value of q (q_0) is a measure of the initial physiological state of the cells. A more stable transformation of q_0 may be defined as

$$h_0 = \ln\left(1 + \frac{1}{q_0}\right) = \mu_{max}\lambda \qquad (3.8)$$

Thus, the final model has four parameters: x_0, the initial cell number; h_0; x_{max}; and μ_{max}. The parameter h_0 describes an interpretation of the lag first formalized by Robinson et al. (1998). Using the terminology of these authors, h_0 may be regarded as the 'work to be done' by the bacterial cells to adapt to their new environment before commencing exponential growth at the rate μ_{max} characteristic of the organism and the environment. The duration of the lag, however, also depends on the rate at which this work is done, which is often assumed to be μ_{max}.

3.2.1.3 Logistic and Linear Functions

Rosso et al. (1996) demonstrated that the logistic model using delay and rupture was a model that, compared to those available at the time, provided very good accuracy using only four descriptive parameters: lag time (t_{lag}), growth rate (μ_{max}), initial population size (N_0), and the maximum population density (N_{max}).

$$\begin{cases} \ln(N) = \ln(N_0), & t \leq lag \\ \ln(N) = \ln(N\max) - \ln\left[1 + \left(\frac{N\max}{N_0} - 1\right)\exp(-\mu\max(tlag - lag))\right], & t > lag \end{cases}$$

$$(3.9)$$

Buchanan et al. (1997) proposed a three-phase linear model that can be described by three phases: lag phase, exponential growth phase, and stationary phase, as follows.

Lag phase:
For

$$t \leq t_{lag}, N_t = N_0 \qquad (3.10)$$

Exponential growth phase:
For

$$t_{lag} < t < t_{max}, N_t = N_0 + \mu\left(t - t_{lag}\right) \qquad (3.11)$$

Stationary phase:
For

$$t > t_{max}, N_t = N_{max} \qquad (3.12)$$

where N_t is the log of the population density at time t (log cfu/ml); N_0 is the log of the initial population density (log cfu/ml); N_{max} is the log of the maximum population density supported by the environment (log cfu/ml); t is the elapsed time; t_{lag} is the time when the lag time ends (h); and μ is the maximum growth rate (log cfu/ml/h).

In this model, the growth rate is always at maximum between the end of the lag phase and the start of the stationary phase while μ is set to zero during both the lag and stationary phases. The lag is divided into two periods: a period for adaptation to the new environment (t_a) and the time for generation of energy to produce biological components needed for cell replication (t_m).

3.2.1.4   Compartmental Models

The McKellar model assumes that bacterial population exists in two 'compartments' or states: growing or nongrowing. All growth was assumed to originate from a small fraction of the total population of cells that are present in the growing compartment at $t = 0$. Subsequent growth is based on the following logistic equation:

$$\frac{dG}{dt} = G \cdot \mu \cdot \left(1 - \frac{G}{N_{max}} \right) \tag{3.13}$$

where G is the number of growing cells in the growing compartment. The majority of cells were considered not to contribute to growth, and remained in the nongrowing compartment, but were included in the total population. Although this is an empirical model, it does account for the observation that the first cells to begin growth dominate growth in liquid culture, and that any cells that subsequently adapt to growth are of minimal importance (McKellar et al. 1997). The model derives from the theory that microbial populations are heterogeneous rather than homogeneous, existing as two populations of cells that behave differently; the sum of the two populations effectively describes the transition from lag to exponential phase, and defines a new parameter G_0, the initial population capable of growing. Reparameterization of the model led to the finding that a relationship existed between μ_{max} and λ as described in the Baranyi model. In fact, Baranyi and Pin (2001) stated that the initial physiological state of the whole population could reside in a small subpopulation. Thus, the McKellar model constitutes a simplified version of the Baranyi model and has the same parameters.

The concept of heterogeneity in cell populations was extended further to the development of a combined discrete–continuous simulation model for microbial growth (McKellar and Knight 2000). At the start of a growth simulation, all the cells were assigned to the nongrowing compartment. A distribution of individual cell lag times was used to generate a series of discrete events in which each cell was transferred from the nongrowing to the growing compartment at a time corresponding to the lag time for that cell. Once in the growing compartment, cells start

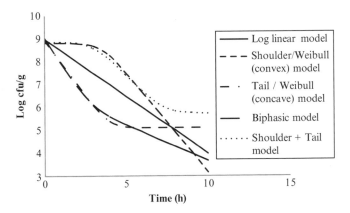

Fig. 3.2 Generic models used for describing the inactivation of pathogens in foods

growing immediately, according to (Eq. (3.13)). The combination of the discrete step with the continuous growth function accurately described the transition from lag to exponential phase.

At the present time it is not possible to select one growth model as the most appropriate representation of bacterial growth. If simple is better, then the three-phase model is probably sufficient to represent fundamental growth parameters accurately (Garthright 1997). The development of more complex models (and subsequently more mechanistic models) will depend on an improved understanding of cell behavior at the physiological level.

3.2.2 Inactivation Models

One of the quantitative microbiology tools for microbial inactivation is a freeware add-in for Microsoft Excel, the so-called GInaFiT (Geeraerd and Van Impe Inactivation Model Fitting Tool), which includes a large variety of primary inactivation models by the using of linear and nonlinear regression approaches. The application, described by Geeraerd et al. (2005), comprises nine model types: (1) classical log-linear curves, (2) curves displaying a so-called shoulder before a log-linear decrease is apparent, (3) curves displaying a so-called tail after a log-linear decrease, (4) survival curves displaying both shoulder and tailing behavior, (5) concave curves, (6) convex curves, (7) convex/concave curves followed by tailing, (8) biphasic inactivation kinetics, and (9) biphasic inactivation kinetics preceded by a shoulder.

The models most used for describing the inactivation of pathogens in foods are presented in Fig. 3.2, and a brief description of the various models is provided next.

3.2.2.1 Bigelow Model (Linear Model)

This model was established to quantify microbial inactivation in the canning industry, assuming first-order kinetics. Apparently, cell death occurs as a result of inactivation of a critical enzyme, and it is commonly stated that this inactivation follows first-order kinetics (although there might be some exceptions to this rule). The model has the following form:

$$\ln N = \ln N_0 - kt \tag{3.14}$$

with N as the number of microorganisms, N_0 as the initial number of microorganisms, and k as the first-order rate constant (seg^{-1}). This equation is then rearranged into

$$\log \frac{N}{N_0} = \log S(t) = -\frac{t}{D} \tag{3.15}$$

where D is the decimal reduction time ($D = 2.303/k$, units in minutes or seconds), and $S(t)$ is the momentary survival ratio.

A drawback of this model is that it is mainly focused on thermal treatments rather than on novel preservation techniques.

When log D-values are plotted against temperature, the reciprocal of the slope is equal to the z-value, which is the temperature increase needed to reduce D by a factor 10, so as to increase the destruction rate by a factor of 10.

$$D = D_{ref} 10^{\left(\frac{T_{ref} - T}{z}\right)} \tag{3.16}$$

$$z = \frac{T_{ref} - T}{\log D - \log D_{ref}} \tag{3.17}$$

where D_{ref} is the D-value at the reference temperature, T_{ref} (the usual T_{ref} is 121.1°C). D-values could be influenced by the type of organism (strain), treatment temperature, physiological state of cells, pH, fat, and a_w content.

The rate constant can also be related to the temperature by the Arrhenius equation:

$$k = k_0 \exp\left(-\frac{Ea}{RT}\right) \tag{3.18}$$

where Ea is the activation energy, k_0 is the collision factor (regression coefficient), and R the universal gas constant, and T is the temperature in Kelvin degrees. The Arrhenius model has been used for describing the kinetic behavior of *Escherichia coli* as a function of temperature, pH, and a_w (Cerf et al. 1996).

One parameter, named the sterilizing value (F), is used to designate the thermal death time, or the time to kill all pathogenic organisms at 121.1°C.

The F value can be defined as the time taken to reduce initial microbial numbers, at a specified temperature, by a particular value, normally a multiple of the D-value for the target organism. The process goal as described by the sterilizing value is given by the expression:

$$F = D(\log N_0 - \log N) \qquad (3.19)$$

For a nonacid food the minimum process must assure safety by destroying any contaminating *Clostridium botulinum*. As suggested by Stumbo et al. (1975), this is considered to accomplish the 12D process, which is a 10^{-12} reduction of the N_0 value.

Because of its broad applicability, the log-linear model is most appropriate to obtain a first impression on the performance of an inactivation process. This characteristic is especially useful for the food industry where elaborate knowledge and the necessary tools for complicated models and generic parameter values are not available (van Asselt and Zwietering 2006).

3.2.2.2 Weibull Model

The Weibull model has been used as a primary thermal inactivation model for vegetative bacteria (van Boekel 2002). This model assumes nonlinearity of semi-logarithmic survivor curves in the inactivation process through considering biological variation with respect to thermal inactivation and is basically a statistical model of distribution of inactivation times. The model is built by two parameters, the scale parameter α (time) and the dimensionless shape parameter β. The logarithm of the scale parameter α depends linearly on temperature; however, this relationship for parameter β is not so well established. The shape parameter accounts for upward concavity of a survival curve ($\beta < 1$), a linear survival curve ($\beta = 1$), and downward concavity ($\beta > 1$). Therefore, if $\beta = 1$ no biological variation is assumed (each cell is equally susceptible to be destroyed).

In terms of a survival curve, the cumulative function is

$$\log S(t) = -\frac{1}{2.303}\left(\frac{t}{\alpha}\right)^{\beta} \qquad (3.20)$$

The Weibull distribution function (which is close related to the gamma, extreme value, and log-normal distributions) is widely used in reliability engineering to describe time to failure in electronic and mechanical systems and is also appropriate for the analysis of survival data, that is, time to failure after the application of stress.

Although the Weibull model is of an empirical nature, a link can be made with physiological effects. $\beta < 1$ indicates that the remaining cells have the ability to adapt to the applied stress, whereas $\beta > 1$ indicates that the remaining cells become increasingly damaged.

This variation in parameters allows the Weibull model to be more flexible than the linear model on the basis of D-values. For instance, cell behavior may be quite different when cells have been adapted to certain stress conditions in foods or have been grown in 'ideal' laboratory conditions.

3.2.2.3 Shoulder/Tail Models

Shoulder/tail models are based upon the existence of a shoulder or lag period (before inactivation) and a tail region (after the inactivation treatment). The first type of deviation is produced when the curve is flat; that is, no change in the number of microorganisms at the beginning of the inactivation treatment. The second type is the tailing of survivor curves, which occurs at the end of the inactivation treatment and is characterized by the culture showing more resistance than would be expected from the previous logarithmic order of destruction.

Stringer et al. (2000) has summarized possible explanations for this deviation from linearity such as variability in heating procedure, use of mixed cultures, and clumping or protective effect of the food matrix or dead cells.

A linear approach was followed by Whiting (1993) for modeling inactivation of *Listeria monocytogenes* and *Salmonella* as a function of temperature and in presence of NaCl, nitrites, and lactic acid.

$$\log N = \begin{cases} \log N_0 & \text{when} \quad 0 < t < tL \\ \log N_0 - \left(\dfrac{1}{D}\right)(t - tL) & \text{when} \quad t > tL \end{cases} \tag{3.21}$$

where N is the number of microorganisms surviving at time t, N_0 is the initial microbial load, tL is the time before inactivation, and D is the D-value.

The model was successfully applied to describe nonthermal inactivation of *L. monocytogenes* as a function of additional preservatives and reduced oxygen (Buchanan and Golden 1995).

Regarding the shoulder region, model fitting is more difficult because a high variability is associated with this parameter. Thus, survival is often described through the time required for a 4-log reduction, $T4D$ (Whiting 1993). This value is calculated as the sum of $tL + 4D$.

Nonlinear approaches normally represent a shoulder/tailing function such as

$$\log \frac{N}{N_0} = -\frac{t^p}{D} \tag{3.22}$$

where p is the power that takes a concave curve when is lower than 1 and a convex (shoulder curve) when is higher than 1.

Other modeling approaches use the logistic functions, in which their inactivation forms are called the Fermi equation. For the quantification of sigmoid decay curves, the following expression is used:

$$\log \frac{N}{N_0} = \log\left[\frac{1 + \exp^{-btL}}{1 + \exp^{b(t-tL)}}\right] \tag{3.23}$$

where N is the number of microorganisms surviving at time t, N_0 is the initial microbial load, b is the maximum specific decay rate, and tL is the time before inactivation.

Geeraerd et al. (2000) developed a shoulder/tail inactivation model considering the physiological state of cells and the residual population density (tail region):

$$N = \left[\left(N_0 - N_{res}\right)\exp(-k_{max}t)\frac{\exp(-k_{max}tL)}{1 + \exp((-k_{max}tL) - 1)\exp(-k_{max}t)} + N_{res}\right] \tag{3.24}$$

where N is the number of microorganisms surviving at time t, N_0 is the initial microbial load, k_{max} is the maximum specific decay rate, tL is the time before inactivation, and N_{res} is the residual population density.

Other inactivation models relied upon mechanistic processes in which predictions can be achieved outside the range of the obtained experimental data. These models have been mainly developed for spore-forming bacteria and for high-temperature treatments (UHT, HTST sterilization). However, one drawback of more complex and mechanistic models lies in the difficulty in being able to successfully develop them, and the primary drawback is the absence of generic parameters.

3.3 Secondary Models

These models predict the changes in the parameters of primary models such as bacterial growth rate and lag time as a function of the intrinsic and extrinsic factors.

Mathematical expressions (secondary models) can be distinguished as two different approaches:

1. The effects of the environmental factors are described simultaneously through a polynomial function; this type of model has probably been the most extensively used within predictive microbiology
2. The environmental factors are individually modeled, and a general model describes the combined effects of the factors; this approach is notably applied in the development of the increasingly popular square root and cardinal parameter-type models

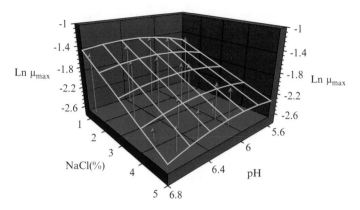

Fig. 3.3 Response surface representing the relationship between the ln of the maximum growth rate (μ_{max}) and pH and NaCl levels

3.3.1 Polynomial Models

Polynomial models (so-called response surface models) are empirical secondary models in which the high degree of fitting and ease of elaboration are their most significant advantages. Generally, second-order polynomial equations are used, including three terms: first order, second order (quadratic), and interaction terms. Several studies have described the relationship between certain environmental factors (temperature, pH, a_w, NaCl, $NaNO_2$, etc.) by means of these types of equations (Lebert et al. 2000; Zurera et al. 2004):

$$y = \beta_0 + \sum_{j=1}^{k} \beta_j X_j + \sum_{j=1}^{k} \beta_{jj} X_j^2 + \sum_{j \neq 1}^{k} \beta_{jl} X_j X_l + \varepsilon \qquad (3.25)$$

where y is the dependent variable (growth rate, lag phase etc.), $\beta_0, \beta_j, \beta_{jj}, y \beta_{jl}$ are the estimated regression coefficients X_j y X_l are the independent variables (environmental factors), and ε is the error term.

Polynomial models are characterized by a high number of parameters, which increase exponentially when increasing the number of factors included in the model. In turn, their ease of implementation and handling allow them to be developed in many computing software and model applications. The main pitfalls of these models are that they are too flexible (i.e., the excessive number of parameters can give wrong estimations because error can be also modeled) and the absence of terms that can explain biological behavior.

Response surface models can be graphically translated, as described in Fig. 3.3, where the relationship between ln μ_{max} and pH and NaCl levels can be seen. Food operators can easily implement or even optimize formulations of novel foods to achieve the necessary combinations of factors that just inhibit microbial development.

3.3.2 Square Root-Type Models

These secondary models were initially proposed by Ratkowsky et al. (1982), who observed a linear relationship between the square root of the maximum growth rate and temperature (at suboptimal conditions for growth):

$$\sqrt{\mu_{max}} = b \cdot (T - T_{min}) \tag{3.26}$$

where T_{min} is the notional minimum temperature below which maximum growth rate is equal to 0 (it ranges between 2°C and 3°C below the observed minimum temperature). T_{min} is generally obtained through a linear regression of the square root of maximum growth rate and temperature.

Later, this model was extended to cover the whole temperature growth range (Ratkowsky et al. 1983):

$$\sqrt{\mu_{max}} = b \cdot (T - T_{min})\{1 - \exp[c(T - T_{max})]\} \tag{3.27}$$

Other adaptations include the effect of alternative environmental factors, such as pH, a_w, or lactic acid (Ross et al. 2003).

For modeling bacterial lag time, alternative expressions are suggested:

$$\sqrt{\frac{1}{t_{lag}}} = b \cdot (T - T_{min}) \tag{3.28}$$

3.3.3 The Gamma Concept and the Cardinal Parameter Model (CPM)

Zwietering et al. (1992) proposed a model called the gamma model describing the growth rate relative to its maximum value at optimal conditions for growth:

$$\mu_{max} = \mu_{opt} \ \gamma(T) \ \gamma(pH) \ \gamma(a_w) \tag{3.29}$$

where μ_{opt} is the growth rate at optimum conditions, and $\gamma(T)$, $\gamma(pH)$, and $\gamma(a_w)$ are the relative effects of temperature, pH, and a_w, respectively. The concept underlying this model (the gamma concept) is based on the following assumptions: (1) the effect of any factor on the growth rate can be described, as a fraction of μ_{opt}, using a function (γ) normalized between 0 (no growth) and 1 (optimum condition for growth); and (2) the environmental factors act independently on the bacterial growth rate. Consequently, the combined effects of the environmental factors can be obtained by multiplying the separate effects of each factor (Eq. (3.28)). Ross and

Dalgaard (2004) considered that although this apparently holds true for growth rates, environmental factors do interact synergistically to govern the biokinetic ranges for each environmental factor.

At optimal conditions for growth, all γ terms are equal to 1 and therefore μ_{max} is equal to μ_{opt}. The γ terms proposed by Zwietering et al. (1992) for the normalized effects of temperature, pH, and water activity are given in Eqs. (3.30), (3.31), and (3.32).

$$\gamma(T) = \left(\frac{T - T_{min}}{T_{opt} - T_{min}} \right)^2 \tag{3.30}$$

$$\gamma(pH) = \frac{pH - pH_{min}}{pH_{opt} - pH_{min}} \tag{3.31}$$

$$\gamma(a_w) = \frac{a_w - a_{wmin}}{1 - a_{wmin}} \tag{3.32}$$

The γ-type terms for pH and lactic acid effects on growth rate were also included in square root-type models by Presser et al. (1997) and Ross et al. (2003).

Introduced by Rosso et al. (1995), the cardinal parameter models (CPMs) were also developed according to the gamma concept. The relative effects of temperature, pH, and a_w on bacterial growth rate are described by a general model called CPM_n:

$$CM_n(X) = \begin{cases} 0, & X \leq X_{min} \\ \dfrac{(X - X_{max})(X - X_{min})}{(X_{opt} - X_{min})^{n-1}\left[(X_{opt} - X_{min})(X - X_{opt}) - (X_{opt} - X_{max})((n-1)X_{opt} + X_{min} - nX)\right]}, & X_{min} < X < X_{max} \\ 0, & X \geq X_{max} \end{cases} \tag{3.33}$$

where X is temperature, pH, or a_w; X_{min} and X_{max} are, respectively, the values of X below and above which no growth occurs; X_{opt} is the value of X at which bacterial growth is optimum; and n is a shape parameter. As for the gamma model of Zwietering et al. (1992), $CM_n(X_{opt})$ is equal to 1, and $CM_n(X_{min})$ and $CM_n(X_{max})$ are equal to 0.

For the effects of temperature and pH, n is set to 2 and 1, respectively (Augustin et al. 2000a, b; Le Marc et al. 2002; Pouillot et al. 2003).

For the effects of a_w, n is set to 2 (Augustin and Carlier 2000a, b). The combined effects of the environmental factors are also obtained by multiplying the relative effects of each factor. Thus, the Cardinal Parameter Model for the effects of temperature, pH, and a_w on μ_{max} can be written as:

$$\mu_{max} = \mu_{opt} \; CM_2(T) \, CM_1(pH) \, CM_2(a_w) \xi(T, pH, a_w) \tag{3.34}$$

or alternatively:

$$\mu_{\max} = \mu_{opt} \quad CM_2(T) \; CM_1(pH) \; CM_1\left(a_w\right) \xi(T, pH, a_w) \qquad (3.35)$$

This model includes an additional term $\xi(T, \mathrm{pH}, a_W)$ to account for the interactions between the environmental factors.

The advantages of the CPMs lie in the lack of structural correlation between parameters and the biological significance of all parameters (Rosso et al. 1995).

Several attempts have been made to include in CPMs the effects of organic acids (Le Marc et al. 2002) or other inhibitory substances (Augustin et al. 2000).

3.3.4 Artificial Neural Networks (ANNs)

Artificial neural networks (ANN) are an artificial intelligence tool that has become increasingly popular in different scientific areas including those concerned with biological systems. They have appeared as an alternative to nonlinear models, including a wide application in several scientific areas. The origins in the predictive microbiology field have arisen thanks in part to their flexibility and the generation of models with a high degree of accuracy to experimental data, in comparison with other regression techniques (Basheer and Hajmeer 2000). Indeed, ANN models have been used for describing growth, inactivation, and probability of growth of microorganisms in different media (Hajmeer et al. 1997). ANN are inspired by the functioning of neurons in human brain. ANN are empirical models able to explain dependency between explanatory variables and response variables irrespective of both the nonlinearity level between variables and independence and normality assumptions, which are important constraints for other regression methods such as LS. The ANN derives arbitrary nonlinear multiparametric discriminant functions directly from experimental data (Almeida 2002).

ANNs constitute empirical approaches able to model (independently of the nonlinearity degree) the existing relationship between a dependent variable (or response variable) and a series of independent factors. It is noted that ANN are not assuming a previous hypothesis of normality and independency between independent factors. One advantage lies in obtaining various multi-equation models with different outputs starting with the same neural network model, thus resulting in a smaller estimation error. In turn, these models present more complexity than linear regression techniques and minor interpretation. Additionally, a lack of mechanistic basis to describe microbial behavior is known.

These models have been described in several studies (Basheer and Hajmeer 2000). Their application in predictive microbiology is found in some related papers such as those by Hajmeer et al. (1997), Geeraerd et al. (1998), Hervás et al. (2001), or García-Gimeno et al. (2002).

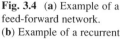

Fig. 3.4 (a) Example of a feed-forward network. (b) Example of a recurrent network

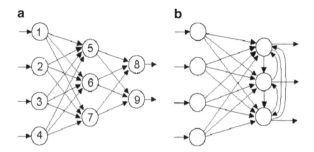

An ANN model is defined as a computational statistical model, with a network-interconnected structure of single components, named neurons or nodes, which are able to make parallel calculations.

The main features characterizing a neural network are the topology, learning mechanism, association between input and output information, and presentation of the generated information.

The topology or architecture of a neural network consists of the organization of neurons inside the network forming different layers or aggregations (Fig. 3.4). In this sense, the main parameters of a given network are as follows:

(a) Number of levels and/or layers
(b) Number of neurons per layer
(c) Connectivity degree between nodes (defined as the number of weights or parameters)
(d) Function types between the different layers, which can be lineal, sigmoid, or hyperbolic.

Distribution of neurons inside the network is achieved by means of forming different layers with a specific number of neurons each. As a function of the neuron position inside the network, three different layers can be differentiated:

(a) Input layer: formed by the independent factors of the model (temperature, pH, a_w, etc.).
(b) Hidden layer: internal layers without direct contact with the external side of the network. Neurons of hidden layers can be interconnected in different ways that conform (together with their number) with the network topology.
(c) Output layer: formed by the dependent variables (μ_{max}, lag, etc.).

In Fig. 3.4 are seen two different forms of neural networks, based on their connectivity. In feed-forward networks, nodes are numbered in such a way that those with a higher number are never connected with those with lower numbers. In the second type, the recurrent network, this numeration method is not followed.

The learning process is defined as a modification in the weights of the network in response to input information, which is equivalent to the adjustment made in the traditional regression models. The learning rule or algorithm most commonly used is the error retro propagation algorithm: it consists of a descendent gradient method to modify the weights of the network connections with the aim of minimizing the error obtained between the desired output values and those obtained by the ANN model.

The type of association between input and output information is represented by a mathematical function. The nomenclature used for presenting this information is 'number of input nodes':'number of hidden nodes' **f**: 'number of output nodes' **t**, where **f** is the function type (sigmoid, linear, or hyperbolic) toward the nodes of the hidden layer and **t** represents the transfer function type toward the nodes of the output layer.

Among the different ANN architectures, one of the most important is the multi-layer perceptron (MLP), defined as a feed-forward ANN model that maps sets of input data onto a set of appropriate output. An MLP consists of multiple layers of nodes in a directed graph, with each layer fully connected to the next one.

As an alternative, different ANN models have been developed, based on radial basis functions, or RBF (Lee and Hou 2002), or general network regression models (Schepers et al. 2000). Product-unit neural networks models (PUNN) are ANN models in which the computing function between nodes located in the hidden layer is a multiplicative type: this allows implementation of higher-degree polynomial equations, as shown by Valero et al. (2007).

The development of an ANN model implies a previous training of the network to optimize the number of nodes in relationship to the variables included in the model. The use of evolutionary algorithms (EA) reduces the probability that the searching process of the solution does not reach the global optimum. EAs are based on a local stochastic searching process, which allows finding a solution in complex matrices.

The general evolutionary process is based on the use of selection, replication, and mutation operators (parametric and structural). Evolution of the network topology corresponds to a local search for the structure of sigmoid and product-base functions that display the best fit to the training data points. Detailed information about the structure of EA is described in Hervás-Martínez et al. (2006).

Application of ANNs could be a sound technique to be applied in predictive microbiology (Jeyamkondan et al. 2001). The higher accuracy and handling of complex datasets allow them to be useful when several variables have to be described. However, the 'black-box' limitations, described in other related studies (Geeraerd et al. 2004), and the lack of interpretability still constitute a drawback for many food microbiologists for efficient application.

Recently, fuzzy computational methods have been created as an alternative to traditional regression models. In this case, observations do not hold a distribution assumption and one cannot assign all the uncertainty of the model to the randomness aspect of variables. ANN and fuzzy intelligent computational methods offer real advantages over conventional modeling, involving the ability to handle large amounts of noisy data from nonlinear and dynamic systems, especially when the underlying physical relationships are not fully understood. Nevertheless, implementation of these combined modeling techniques often requires expert knowledge and complex datasets.

3.3.5 Secondary Inactivation Models

Although the effect of some environmental characteristics of heating treatments, such as pH or a_w content, upon the heat resistance of bacteria has been recognized for a number of decades, very few attempts were made to develop secondary models to quantify these effects.

Some complete quadratic models have been still recently developed to describe the heat resistance of spores (Juneja et al. 1995): such models are overly parameterized. Fernández et al. (1996) studied combined effects of temperature and pH on the heat resistance of *Bacillus stearothermophilus* and *Clostridium sporogenes*. They proposed two alternative models: a complete quadratic equation

$$\log k = a_0 + a_1 T + a_2 \text{pH} + a_3 T\text{pH} + a_4 T^2 + a_5 \text{pH}^2 \tag{3.36}$$

and a simple linear equation

$$\log k = C_0 + C_1 T + C_2 \text{pH} \tag{3.37}$$

Davey (1993) added, to the logarithmic shape of the Arrhenius equation, a pH and a squared pH term to describe the combined effects of temperature and pH on the destruction rate of *C. botulinum* from data from the literature. They also obtained the following reduced polynomial equation:

$$Lnk = C_0 + \frac{C_1}{T} + C_2 \text{pH} + C_3 \text{pH}^2 \tag{3.38}$$

where T is the absolute temperature. With respect to a complete quadratic model (Eq. (3.37)), the Davey model presents the advantage of being more parsimonious (four parameters instead of six) for a satisfactory degree of goodness of fit. However, as in the case of any polynomial model, coefficients remain without any biological meaning.

Evidence for an interacting effect between temperature and pH are referred in the study of Fernández et al. (1996) for *C. sporogenes*. The model, subsequently developed by Mafart and Leguérinel (1998), considers an additive effect on log scale as follows:

$$\log D = \log D^* - \frac{T - T^*}{z_T} - \frac{(pH - pH^*)^2}{z_{pH}^2} \tag{3.39}$$

where T^* is the reference temperature, pH* is the pH of the maximum heat resistance (7.0 for spores), and z_T is the commonly used thermal z-value and z_{pH} is the distance from pH to pH*, which leads to a tenfold reduction of the decimal reduction time, and D^* is the D-value at T^* and pH*.

The a_w was taken into account for the first time by Reichart (1994) who derived a semi-empirical model for the death rate of *E. coli*. A five parameter model was proposed by Cerf et al. (1996) including an extension of the Davey's model with the a_w term:

$$Lnk_{max} = C_0 + \frac{C_1}{T} + C_2 pH + C_3 pH^2 + C_4 a_w^2 \qquad (3.40)$$

where C_0, C_1, C_2, C_3, and C_4 are empirical coefficients without biological significance.

Leguérinel and Mafart (1998) modified the Bigelow model to include an extension with the a_w term. The following relationship emerged:

$$\log D = \log D^* - \frac{T - T^*}{z_T} - \frac{(pH - pH^*)^2}{z_{pH}^2} - \frac{a_w - 1}{z_{aw}} \qquad (3.41)$$

where z_{aw} is the distance of a_w from 1, which leads to a tenfold increase of the D-value.

Published models for other vegetative microorganisms are available, such as the study of Blackburn et al. (1997), who developed a secondary inactivation model for *Salmonella enteritidis* and *E. coli* O157:H7. Alternative secondary models for *L. monocytogenes*, taking into account other factors such as type of organic acid and type of solute to reduce a_w values, were subsequently developed (Miller et al. 2009).

3.4 Predictive Modeling at Dynamic Conditions

In the previous chapters, an overall description of the use of predictive models was provided. Some potential applications were described such as prediction of food safety and shelf life; or the establishment of critical control points in a HACCP system.

However, in the origins of predictive microbiology, scientific studies were biased to the development of predictive models at static conditions. The use of sigmoid functions provided a good description of microbial growth under nonvarying environmental conditions. It is worthy to say that, in a real food process, many factors are influencing microbial behavior and they do not maintain static values along the food chain. Normally, the variation in these factors is normally considered in the primary models and further included into the development of secondary models.

To take into account the real food conditions, models explaining microbial dynamics are needed. Indeed, when extrapolating predictions from predictive models made on static conditions to more realistic dynamic conditions, predictions may fail to describe accurately microbial evolution.

There are different factors that can be added as building blocks into an elementary dynamic model, namely, fluctuating environmental conditions, variability in the physiological state of cells, interactions, and production of metabolites affecting microbial growth.

These factors can be incorporated into a model equation to obtain accurate predictions within the food industry. However, mathematical complexity should be embedded into user-friendly tools to improve their industrial applicability.

The elemental equation that describes a dynamic model can be expressed as follows:

$$\frac{dN_i(t)}{dt} = m_i \cdot N_i(t),\ N_j(t),\ fact(t),\ chem(t),\ phys(t) \cdot N_i(t) \tag{3.42}$$

where $N_i(t)$ represents the cell density of microbial species, μ_i (h^{-1}) is the specific rate resulting from the interactions within (N_i) and between (N_j) microbial species, physicochemical environmental conditions (fact) physiological state of cells (phys), and production of metabolites (chem). Multiplication of the specific growth/inactivation rates at fluctuating conditions are time dependent. Total growth/inactivation will be obtained at the final time of the dynamic profile if $\mu_i > 0$ or $\mu_i < 0$, respectively. To describe the time-dependent evolution of each factor included in the model, additional differential equations must be constructed for each of them.

There are several dynamic models in predictive microbiology (Baranyi et al. 1993; Hills and Mackey 1995; McKellar 2001). One of the most commonly used is the Baranyi and Roberts model, as previously shown in Eqs. (3.6, 3.7, 3.8). This model can be adapted to be used for predicting dynamic growth and inactivation in specific food matrices, as shown in Pin et al. (2011) or Psomas et al. (2011).

As already mentioned, environmental conditions in food products are time dependent. In these cases, model predictions can be generated by combining a dynamic primary model with a secondary model relating the primary parameters (e.g., μ_{max}, lag) with environmental conditions. The final equation will assume that the specific rates will change according to the fluctuation of the environmental factors.

The theory of the cumulative lag was suggested by Koutsoumanis (2001) to model microbial growth at dynamic conditions: this was based on a theoretical expression of lag time, being the ratio between the amount of work (W_n) that a cell needs to do to adapt to its new environment and the rate (R) at which it is able to do that work:

$$t_{lag} = W_n/R \tag{3.43}$$

The work needed (W_n) can be any biosynthetic or homeostatic process that the cell needs to do after its transition from environment E_1 to environment E_2. If one takes the time of transition from E_1 to E_2 as time zero, the 'work accomplished' by the cell accumulates with time, at a rate that depends on the storage temperature. The lag time can be calculated as the time required for the 'work accomplished' to reach W_n.

Consequently, the lag time at fluctuating conditions can be calculated as follows:

$$\int_0^{t_{lag}} dW = W_n \tag{3.44}$$

$$\int_0^{t_{lag}} R[T(t)]dt = W_n \tag{3.45}$$

$$\int_0^{t_{lag}} \frac{1}{L[T(t)]}dt = 1 \tag{3.46}$$

$$\int_0^{t_{lag}} [T(t) - T_{min}]^2 dt = \frac{1}{b^2} \tag{3.47}$$

where t is the time, R is the rate of 'work accomplished' for an assumed constant temperature time interval dt, L is the lag time corresponding to the temperature of this interval, t_{lag} is the total lag time, T is the temperature, and T_{min} and b are the square-root model parameters of lag time (see (Eq. 3.28)).

Other straight forward approaches are based on the use of the term time-temperature-equivalent (TTE), which is linked to the time, temperature, and type of microorganism. This approach also considers as starting point the square-root secondary model (Rosset et al. 2004):

$$TTE = \sum_{i=1}^{n} t_i \cdot (T_i - T_{min})^2 \tag{3.48}$$

where t_i is the specific time duration between the step i and $i + 1$; T_i is the temperature (°C) assumed constant between i and $i + 1$, and T_{min} is the theoretical minimum temperature required for growth described by the square root model.

According to a specific time–temperature profile, the 'effective static temperature' (T_{eff}) can be estimated for the total duration of the profile, t_{tot}:

$$T_{eff} = \sqrt{\frac{\sum_{i=1}^{n} t_i \cdot (T_i - T_{min})^2}{t_{tot}}} + T_{min} \tag{3.49}$$

Concerning microbial inactivation, to predict the effect of non-log-linear microbial behavior, Geeraerd et al. (2000) described the following functions to model the shoulder and tailing effects:

$$\frac{dN(t)}{dt} = -\left[\frac{1}{1 + Cc(t)}\right] \cdot k_{max} \cdot \left[1 - \frac{N_{res}}{N(t)}\right] \cdot N(t) \tag{3.50}$$

$$\frac{dCc(t)}{dt} = -k_{max} \cdot Cc(t) \qquad\qquad (3.51)$$

Before first-order inactivation takes place, some critical protective component, Cc, must be inactivated. The shoulder is obtained by applying a Michaelis–Menten function, which can be interpreted as the physiological state of the population in the context of inactivation. Afterwards, microbial inactivation is assumed to follow a linear approach, reaching the maximum inactivation rate, k_{max}. The tailing phenomenon is explained by the existence of a residual population, N_{res}, which can vary when modeling nonthermal or sequences of inactivation treatments (Shadbolt et al. 2001).

Other dynamic models consider the cumulative effect in the microbial inhibition of food-borne pathogens given by the addition of protective cultures, building as much as possible on already-developed model structures or the Jameson effect (Delignette-Müller et al. 2006). This point will be further described in this brief ('Between-species interaction models').

3.5 Developing Predictive Models: Fitting Methods

3.5.1 Selection of the Model

In predictive microbiology, most models are devoid of biological basis, and therefore are built on the basis of an empirical approach. Hence, mathematical models are derived by searching for the mathematical function(s) that adequately fit observed values. Note that growth data (concentration vs. time) are used to derive primary models whereas kinetic parameters from primary models, that is, lag time, maximum growth rate, and maximum population density, are used to obtain secondary models. In both cases, the selection of the mathematical function(s) to be fitted should be based on a reasonable concordance between the theoretical basis and characteristics of the model and the observed behavior of the variable. To illustrate this, we use a simplified example based on growth data of *E. coli* O157:H7 in culture broth measured over time at different incubation temperatures between 5°C and 25°C. In this case, we focus on the secondary model. If the estimated values of $\sqrt{\mu}$ for the growth data are plotted against the corresponding incubation temperatures, a roughly straight-line relationship may be observed between both variables as evidenced by the data points represented in Fig. 3.5. Thus, predictive microbiology practitioners should look for mathematical functions featured by a straight-line pattern. In this case, based on available secondary models presented in Sect. 3.3, one finds that the square root model could be applied to fit the observed growth for the suboptimal temperature range (Ratkowsky 1983).

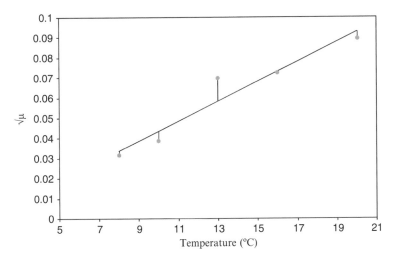

Fig. 3.5 A Ratkowsky's model (*straight line*) for suboptimal temperatures fitted to square root-transformed maximum growth rate values. *Vertical lines* between data points and the *straight line* represent the residuals of the fitted Ratkowsky's model

3.5.2   Fitting Methods

Regression methods are applied to estimate the function's parameters that best describe the observations. The regression method most widely used to fit mathematical functions to data is the least squares method (LS). This method consists of deriving the parameter's values that minimize the sum of the squares of the differences between observed values and those predicted by the fitted model, which are so-called residuals. The residuals represent the distances between data points and the best-fit model and are visualized as vertical lines, as shown in Fig. 3.5. Good performance of the LS method greatly relies on the homoscedasticity assumption, which means that the different data sets of the response variable should show equal variance. When the data present unequal variance, parameter estimates may improve using weighted least squares (WLS). The principle here is to assign to each observation a weight that reflects the error of the measurement. In general, the weight w_i, assigned to the ith observation, will be a function of the variance of this observation. Despite its popularity and versatility, LSM has its problems. Probably, the most important drawback of LSM is its high sensitivity to outliers (i.e., extreme observations) (Abdi 2007).

3.5.2.1   Nonlinear Versus Linear Regression

As commented in earlier sections, mathematical functions can be classified into linear and nonlinear functions. Nonlinear functions are characterized by the fact that the mathematical equation depends nonlinearly on one or more unknown parameters whereas linear regression displays a linear relationship between parameters and variables (Smyth et al. 2002). This aspect greatly determines the type of regression method to be applied. In the case of linear regression, the ordinary LS methods can be efficiently applied to derive the parameter values. However, in nonlinear functions, sometimes the derivative functions intended to estimate the parameter values minimizing LS cannot be solved, which means they do not have an explicit form to be applied. In these cases, a nonlinear regression method should be used. The LS method for nonlinear regression is often based on iterative methods. These methods search in a stepwise fashion for the best values of the estimate. Often they proceed by using at each step a linear approximation of the function and refine this approximation by successive corrections. The techniques utilized are known as gradient descent and Gauss–Newton approximations: they correspond to nonlinear least squares approximation in numerical analysis and nonlinear regression in statistics.

3.5.3 Implications of the Error Term and Variable Transformations

As pointed out in Chap. 1, models possess a deterministic and stochastic component. The deterministic part describes the mathematical relationship existing among variables, whereas the stochastic part accounts for the variation found in the response variable (i.e., dependent variable), which cannot be explained by the deterministic model. The stochastic component is represented by the error term (ε) in the model expression. The optimality of the LS method depends on the error or stochastic component being independently, identically, and normally distributed (i.e. normal distribution). When the property of independence of errors is not met, that is, errors are correlated, the solution from the ordinary LS method is not optimal. To overcome this problem, detailed knowledge about the behavior of the correlated errors is necessary, together with the application of specific algorithms such as time-series models or generalized least squares and other nonparametric methods (Chatterjee and Hadi 2006). In relationship to the other two assumptions, nonidentical variance at different values of the response variable can be solved by applying a WLS as already mentioned, besides which the normal distribution assumption is not always determinant. However, modelers usually apply specific mathematical transformations to the response variable to homogenize variance and make the error normally distributed.

Taking logarithms in both sides of the mathematical expression can linearize a function. By linearizing a function enables application of a linear regression analysis and in consequence an ordinary LS method is suitable. However, this procedure can affect the error behavior, which should be considered to better fit the model to data. For example, consider the following exponential model:

$$Y = \alpha e^{\beta X} + \varepsilon \tag{3.52}$$

If neperian logarithms are taken, the expression can be written as

$$\ln(Y) = \ln(\alpha) + \beta X + \varepsilon'$$ (3.53)

where ε' would be distributed differently from ε. In this respect, McMeekin et al. (1993a) indicated that if ε were independent, identical, and normally distributed, the logarithmic transformation would give rise to ε' showing a smaller variance when Y is large in comparison to when Y is small. Therefore, these authors recommended applying a WLS method to handle the unequal variance of the transformed response variable, $\ln(Y)$.

In predictive microbiology the response variable in secondary models, that is, the lag time, maximum growth rate, and maximum population density, are also transformed to obtain better fitting according to the stochastic assumption considered in each case and model. A particular case is the lag time whose value is more variable at low temperatures far from the optimal temperature. In those cases, the stochastic assumption in the model is quite determinant and special precaution should be taken to choose the most optimal model to be fitted. The most used transformations for this kinetic parameter are based on the log assumption (i.e. $\ln(1/\lambda)$) and square-root assumption (i.e. $\sqrt{1/\lambda}$) assumption as reflected by Ratkowsky (2004). Concerning the maximum growth rate, many works have successfully used the root-square transformation (i.e., $\sqrt{\mu}$) and log-transformation (e.g., $\ln(\mu)$) for this parameter (Ratkowsky 2004, 1983; McMeekin et al. 2002).

3.5.4 Goodness-of-Fit Indexes

Different statistical goodness-of-fit indexes are available to assess if the mathematical function fits well to the data points after a mathematical function is fitted by a regression method. However, not all indexes are optimal to carry out this task, and caution should be taken to select the suitable one. The selection of the statistical index should be made accordingly to the type of function and applied fitting procedure. One important issue is the linearity or nonlinearity of the model, which can determine the type of goodness-of-fit index to be used.

Before proceeding to examine goodness-of-fit indexes, predictive microbiology practitioners should first analyze standardized residuals (i.e., the standardized distance between data point and fitted model) (McMeekin et al. 1993a). Normality assumption for the error term can be ascertained by testing normality in associated residuals based on normal probability plots. Furthermore, linearity and homoscedasticity assumptions can also be evaluated by looking at the standardized residuals with the use of scatter plots. If both assumptions are true, residuals should vary randomly around zero and the spread of the residuals should be about the same throughout the plot. As commented previously, one remarkable weakness of the least squares regression method is its sensitivity to outliers, which can be also detected by looking at residuals.

The root mean square error (RMSE) is probably the most common index to test the goodness of fit of models to the data. Its simplicity and easy interpretation make it suitable for a first approach to the fitted model. Also, RMSE is a valid index for both linear and nonlinear mathematical functions (Ratkowsky 1983, 2004). A low RMSE value indicates a good fitting to data as a result of the closeness of the data points to the fitted model. In turn, a high RMSE value signals that the data points are far from the fitted models, that is, a poor fit to the data.

$$RMSE = \sqrt{\frac{\sum_{i=1}^{n} (Y_i - \hat{Y}_i)^2}{n}} \tag{3.54}$$

here Y_i corresponds to observed value; $\hat{Y}_i$ is the predicted value and n is the number of data points or observations.

One drawback of using RMSE is that it is not a standardized measure and depends on the magnitude of the data value, so that models from different data sets cannot be compared if, for example, units are different, such as meters and KPa. To overcome this issue, an scaled index can be used such as the mean relative error (MRE) in which the magnitude of the data is taken into consideration, including a magnitude term in the denominator (Karadavut et al. 2010).

$$MRE = \frac{1}{n} \left(\frac{\sum_{i=1}^{n} |Y_i - \hat{Y}_i|}{Y_{max} - Y_{min}} \right) \tag{3.55}$$

where Y_{max} and Y_{min} correspond to the maximum and minimum value of observations, respectively.

In general, linear models can be adequately assessed by using the coefficient of determination (R^2), which informs about the fraction of total data point variation explained by the fitted model. This is mathematically defined by the ratio of the variation explained by the fitted mathematical function to the data point variation.

$$R^2 = 1 - \frac{SS_{reg}}{SS_{total}} \tag{3.56}$$

where SS_{reg} corresponds to the sum of squares result from the fitting process, while SS_{total} is the sum of squares of point data or observations.

In spite of being a common practice, R^2 or its adjusted version (adjusted-R^2) should be not applied to evaluate fittings of nonlinear models because some serious limitations are derived from this application. For further information and detail about this issue, we recommend reading Ratkowsky (2004).

To enable comparison between different models, two indexes are mostly used in predictive microbiology. The first is the F-ratio, which can be applied for models with the same number of regression parameters or a different number, if models are nested, meaning that one can be mathematically derived from the other (Zwietering et al. 1990; Wijtzes et al. 1993; Pin et al. 2000; Silva et al. 2010). The F-ratio expression used for the latter case can be written as

$$F = \frac{(SS_1 - SS_2)/(df_1 - df_2)}{SS_2/df_2} \tag{3.57}$$

where SS_1 and SS_2 correspond to the sum of squares resulting from the fitting process for models 1 and 2, respectively, while df_1 and df_2 define the degree of freedom in both models.

A second comparison index used for comparing predictive models is the so-called Akaike's information criterion (AIC), which may be given in its corrected version, corrected Akaike's information criterion (AICc) (Huang et al. 2011) This parameter is particularly suitable to compare nonnested models with a different number of parameters (Akaike 1974). In other words, AICc determines the model with the fewest parameters that still provides an adequate fit to the data. According to AICc, the most adequate model is the one with the lowest AICc value.

3.6 Model Validation

Validation is an essential step in the modeling process. Models cannot be applied if a validation process is not previously accomplished, which typically consists of confirming the predictions experimentally by using any quantitative method (Dym 2004).

In predictive microbiology, experimental analysis of growth in food is the basis of model validation (challenge tests): experimental growth data (i.e., observations) are compared with the model predictions (Gibson et al. 1988; Sutherland and Bayliss 1994). Although validation should be performed on food, in many cases, because of the economic cost derived from challenge tests, data from scientific literature or artificial media are also used for validation. When validation is performed with data sets taken from the same experimental conditions as those used to elaborate the model, validation is the so-called internal validation and aims at determining if the model can sufficiently describe the experimental data. Some authors have carried out internal validation studies obtaining good results (García-Gimeno et al. 2002; Zurera et al. 2004). The external validation is based on the comparison between predictions and independent data sets, that is, either observations obtained from challenge test (Whiting and Buchanan 1994b; te Giffel and Zwietering 1999; Ross et al. 2000) or data taken from scientific literature (Fernandez et al. 1997; McClure et al.1997).

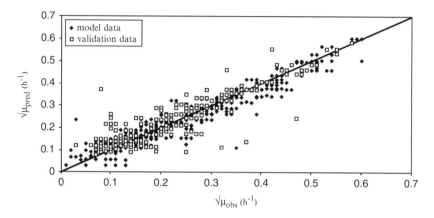

Fig. 3.6 An example of graphical validation in which point data corresponding to model and validation data are situated above and below the equivalence line, which indicates if predictions are fail-safe or fail-dangerous, respectively

In predictive microbiology, diverse methods have been used to compare the goodness of fit of models to experimental data used for model elaboration, and also methods have been used to assess the acceptability of model predictions in relation to the inherent error in experimental data (Zwietering et al. 1994). Nerbrink et al. (1999) graphically compared the observed values of maximum growth rate and generation time found in literature for the same experimental conditions. By means of graphical representation, modelers can easily distinguish between those predictions overestimating and underestimating bacterial growth. Modelers usually classify models into fail-safe models, if predictions overestimate growth, and fail-dangerous models if predictions underestimate growth. In the case of pathogenic bacteria, a fail-safe model is preferred because decisions based on these predictions are more conservative and therefore safer from the public health aspect. Figure 3.6 is a typical example of graphical validation in which fail-safe predictions are situated above the equivalence line, in the left-hand side of the graph. The equivalence line represents for perfect concordance between predicted and observed values. The point data found below the equivalence line, on the right-hand side of the graph, represents for fail-dangerous predictions.

The proximity of observations to the fitted model is assessed by the mean square error (MSE) and the coefficient of determination (R^2), which have been widely applied for validation purposes in food. Another measure of accuracy used for predictive models was introduced by McClure et al. (1993), who compared models based on the sum of the squares of the difference between the natural logarithm of the observed (*gobs*) and predicted (*gpred*) values for generation time:

$$MSE = \Sigma [\ln(gobs) - \ln(gpred)]^2 \tag{3.58}$$

The observed response will be more precise as the value become smaller. By transforming the foregoing equation, the root mean square error (RMSE) is calculated:

$$\text{RMSE} = \sqrt{\frac{\sum_{i=1}^{n} \left(\mu_{\text{obs}} - \mu_{\text{pred}}\right)^2}{n}} \tag{3.59}$$

where μ_{pred} y μ_{obs} corresponds to the maximum growth rate observed and predicted by the model, respectively, and n is the total number of data.

The RMSE is a measure of the residual variability, which cannot be explained by controlled changes in the environmental factors (i.e., explanatory or independent variables) such as temperature and pH. This residual variability may be caused by various elements, including natural variability and certain systematic error. A low RMSE value means better adequacy of the model to describe data (Sutherland et al. 1994). However, if the magnitude of observed data is large, the RMSE will increase proportionally because of the residual variability is higher.

The standard error of prediction (SEP), an index similar to RMSE but using relative terms, which means that its value does not depend on the magnitude of the observed data (Hervás et al. 2001), is defined as

$$\text{SEP} = \frac{100}{\bar{\mu}_{\text{obs}}} \sqrt{\frac{\sum_{i=1}^{n} \left(\mu_{\text{obs}} - \mu_{\text{pred}}\right)^2}{n}} \tag{3.60}$$

Ross (1996) developed two statistical indexes to assess the prediction capacity of models in a simple manner. These indexes are the so-called bias factor (B_f) and accuracy factor (A_f), which give a good estimation of the reliability of models and are written as

$$B_f = 10^{\left[\sum \log(gobs/gpred)/n\right]} \tag{3.61}$$

$$A_f = 10^{\left[\sum |\log(gobs/gpred)|/n\right]} \tag{3.62}$$

The main purpose of these indexes was to enable comparison between predictions and independent observations, not being used for the model elaboration and obtained from food (i.e., external validation), thereby assessing the capacity of models for application in real conditions (Baranyi et al. 1999). The bias factor is an overall average of the ratio of discrete model predictions to observations and assesses whether the model is fail-safe, fail-dangerous, or perfect. A value of 1 means that observations are equally distributed above and below predictions producing a perfect concordance, values $<$ 1 mean a fail-dangerous model, and values $>$ 1 indicate a fail-safe model. The acceptable bias factor value for a predictive model can be 0.75–1.25. The accuracy factor is similar to the bias factor, except that it is the absolute value of the ratio of predictions to observation, thus providing how close predictions are to observations. A value of 1 indicates perfect coincidence, on average, between both predictions and observations, whereas a value of 2 would mean that, on average, predictions are two factors of difference with respect to observations. These validation indexes have been widely used in validation studies (Nerbrink et al. 1999; te Giffel and Zwietering 1999; Ross et al. 2000; Devlieghere et al. 2001).

It is important to note that model users should know the application range or model domain before application as well as the reliability limits that were given by the validation index. When validation indexes are studied, prediction should be analyzed concerning if they are fail-safe or fail-dangerous. Although a certain safety margin is required, it is suggested that predictions are as close as possible to real conditions.

The B_f and A_f were modified by Baranyi et al. (1999) on the basis of the difference of the mean of squared values. The most generic form is represented in the following equation, when models f and g are compared:

$$B_f = \exp\left[\frac{\int_R (\ln f(x_1...x_n) - \ln g(x_1...x_n))dx_1...dx_n}{V(R)}\right] \qquad (3.63)$$

$$A_f = \exp\left[\frac{\int_R (\ln f(x_1...x_n) - \ln g(x_1...x_n))^2 dx_1...dx_n}{V(R)}\right] \qquad (3.64)$$

where $f(x_1...x_n)$ are the predicted values of the maximum growth rate from the function f composed of n environmental factors x; $g(x_1...x_n)$ are the predicted values of the maximum growth rate from the function g composed of n environmental factors x; and $V(R)$ is the volume of the region, that is, if R encompasses the temperature interval (T_2-T_1), then $V(R) = T_2-T_1$.

Campos et al. (2005) proposed a new validation index in terms of microbial counts in contrast to bias factor and accuracy factors that is based on estimations from secondary models, that is, maximum growth rate and lag time, and not log-counts. The robustness index (RI) was defined as the ratio of the standard error of prediction to the standard error of calibration based on log-counts at specific time intervals and for different environmental conditions. The standard error of calibration and standard error of prediction are the root mean square errors estimated from the original and independent data sets, respectively. When RI results in values lower than or close to 1, it means that prediction errors calculated based on independent data are within the expected error of the model, which is defined by the standard error of calibration. Because RI does not provide information regarding whether the model is fail-safe or fail-dangerous as the B_f does, these authors proposed the mean relative error (RE), which is used with the *RI* to provide this information. The equation for RE is written as follows:

$$RE = \frac{\sum\limits_{i=1}^{n} \left(\dfrac{cobs_i - cpred_i}{cpred_i}\right)^2}{n} \tag{3.65}$$

here $cobs_i$ and $cpred_i$ correspond to observed concentration values belonging to an independent data sets and concentration values predicted by the model, respectively expressed in predicted value log(cfu/ml or g), while n specifies the number of observed data from the different data sets.

Chapter 4
Other Models and Modeling Approaches

Abstract Predictive models have been initially focused on the estimation of kinetic parameters, as described in the preceding chapter. However, other modeling approaches are often requested, especially when considering the transmission of a pathogen along the food chain or the probability that this pathogen can grow or survive at certain environmental conditions. This is the underlying reason why transfer and growth/no growth models presented a relevant development in predictive microbiology. These models can be effectively applied when presence/absence data are required, or in specific food processes. Alternatively, survival and transmission of microorganisms through food contact surfaces, environment and between different foods can be also estimated. Additional advantages, such as the extent of shelf life, or the effect of novel preservation methods in minimally processed foods provide a wider application of predictive microbiology. Bacterial transfer models and growth/no growth models are described in this chapter.

Keywords Transfer rate • Cross-contamination • Working surface • Probability of growth • Logit P • Minimum convex polyhedron

4.1 Bacterial Transfer Models

Traditionally, predictive microbiology has focused on kinetic models intended to reflect changes in concentration over time as consequences resulting from physiological activity of cells. During the past few years, the need of describing how microorganisms are transmitted throughout the food chain has led microbiologists to look at other bacterial processes than growth and death. Cross-contamination is reported to be an important factor strongly linked to food-borne disease outbreaks and food spoilage. Hence, some risk assessment studies have pointed out that although there are sufficient kinetic models, some gaps remains to be filled to perform a complete quantitative risk assessment.

F. Pérez-Rodríguez and A. Valero, *Predictive Microbiology in Foods*,
SpringerBriefs in Food, Health, and Nutrition 5, DOI 10.1007/978-1-4614-5520-2_4,
© Fernando Pérez-Rodríguez and Antonio Valero 2013

Cross-contamination refers to the indirect and direct microorganism transfer from a contaminated food surface to other recipient food surfaces in food-related environments. Basically, transfer can occur in many situations and scenarios, for examples, after an inactivation process, which is so-called recontamination, or during food preparation at the retail or consumption stage. The microorganism transfer is a phenomenon rather affecting the number of contaminated food samples in a lot than the concentration levels because bacterial transfer often occurs at low numbers. Nevertheless, as pointed out by Roberts (1990), food-borne outbreaks are often originated by an initial cross-contamination resulting in a contaminated food. Then, careless handing of temperatures increases the low initial concentration up to risk levels, leading to the outbreak.

When concerned with microorganism transfer to foods, a general division can be established as function of the type of microorganism, that is, bacteria and virus. It is evident that during the past few years, food-borne outbreaks derived from pathogenic viruses have increased, and food safety national authorities and the food industry have showed major concern. Virus transfer is particularly important because virus transmission only relies on cross-contamination or transfers from environment to foods because viruses are not able to grow. Nevertheless, few studies have been conducted with virus in food-related environments and no predictive models have yet been derived. Although only recently has bacterial transfer been considered as an important area to be modeled, several studies have attempted to give insight in the transfer process to provide more reliable models and predictions.

As the variety of bacterial transfer being able to occur in food-related environments is ample, scientists working in this area usually classify transfer into three types (Pérez-Rodríguez et al. 2008): air-to-food transfer, surface-to-food in liquids transfer, and surface-to-food by contact transfer.

4.1.1 Air-to-Food Transfer

The impact of bacteria transfer from the air to foods in public health may be considered negligible as there is no epidemiological evidence of food-borne outbreaks linked to bacteria transmission through air. Furthermore, most bacteria are very sensitive to hydric stress, showing reduced viability in the air. Still, exceptions can be observed, such as *Staphylococcus aureus*, a bacterium showing a special persistence in air, and spores and molds, which can remain viable in the air for long periods of time. The air-to-food transfer gains relevance in those foods exposed to air or that use air for food preparation, such as ice cream and instant powders (den Aantrekker et al. 2003b).

Mathematical models have been proposed to represent bacteria transmission from air to foods based on an empirical approach using a quadratic equation relating number of bacteria in the air and bacterial concentration in the food. Also, a predictive model can be obtained using the settling velocity of particles because bacteria are expected to be in the dust and other particles suspended in the air. This model can be easily described by the following equation (Whyte 1986):

$$T = v \cdot C \cdot A \cdot t \tag{4.1}$$

where T = number of bacteria transferred to the food (colony-forming units, cfu); v = the settling velocity of particles (m/h); C = bacterial concentration in the air (cfu/cm^3); A = food area (cm^2); and t = exposure time (h)

The value of v can be calculated by dividing the rate of sedimentation of cells (cfu/m^2·h) by the concentration of cells in the air (cfu/cm^3). Those values can be obtained by experimentation, using an air sampler for bacterial concentration in the air and the Petri dish method to measure the sedimentation rate.

An more extensive review of air-to-food transfer models was carried out by den Aantrekker et al. (2003b) in which probabilistic models describing spores or dioxin-like compounds transfer during a rain event in plants were explained. Although these models were developed for other situations than foods, authors stated that the mechanism is basically the same as bacterial transfer in food-related environments via aerosol or splash formed by spraying contaminated floors.

4.1.2   Surface-to-Food Transfer in Liquids

This type of transfer refers to transfer events in pipeline systems or food liquid tanks that typically occur in the beverage food industry (Pérez-Rodríguez et al. 2008). Mainly, the process starts when biofilms are formed on some specific zones in pipes or tanks. Biofilms are agglomerations of bacteria and organic matter. When biofilm is formed, this may become a microbial contamination source, enabling bacterial transfer from the inner surface of the pipe to the food fluid. When disinfection fails to remove biofilms, it can lead to a significant and continuous bacterial flux to the food liquid flow, resulting in a large number of contaminated food products; hence, this is a serious concern for food industries. Many biofilm models have been developed for predicting microbial contamination in water distribution systems (McBain 2009). They are based on a mechanistic approach, which basically consists of a set of differential equations describing attachment, growth, and detachment of cells in the liquid and solid phases (den Aantrekker et al. 2003b). These models can describe cross-contamination in aqueous foods. The conceptual representation of mathematical models describing the liquid and solid phase is as follows.

Liquid-phase model:

$$\frac{\partial N_L}{\partial t} = \text{Input} + \text{desorption } (Kd) + \text{growth } (\mu) - \text{absorption } (Ka) - \text{output} \tag{4.2}$$

Solid-phase model:

$$\frac{\partial N_S}{\partial t} = \text{Absorption } (Ka) + \text{growth } (\mu) + \text{ desorption } (Kd) \qquad (4.3)$$

where Ka is the absorption constant standing for the capacity of attachment of cells from liquid to solid (e.g., inner surface of the pipe), μ is the maximum growth rate on the inner surface, and Kd is the desorption constant describing the capacity of releasing from the solid to the liquid.

4.1.3 Surface-to-Food Transfer by Contact

The most common bacterial transfer occurring in food-related environments consists of two surfaces coming in contact in which at least one surface is contaminated, that is, the donor surface, which, by contact, transfers cells to other surfaces, named the recipient surface. A bacterial transfer by contact often involves a chain of events in which cells are transferred between surfaces until reaching the food. For instance, transfer from raw meat to a working surface and then from this surface to the final food (e.g., ready-to-eat food) when handled on the surface. In those cases, transfer can occur in different directions and different contamination vectors can be interacting at the same time. For instance, in the foregoing example, knife and worker hands can be also involved in the bacterial transfer to the final food. To model this phenomenon, a simple concept is used so-called transfer rate or coefficient, which stands for the proportion of cells transferred from the donor surface to the recipient surface (Eq. (4.4)):

$$\text{Tr}(\%) = \frac{\text{cfu on recipient surface}}{\text{cfu on donor surface}} \times 100 \qquad (4.4)$$

here, the numerator represents the number of cells (cfu) on the recipient surface after contact and in the denominator the number of cells (cfu) before contact in the donor surfaces.

The transfer rates are estimated by using laboratory models or by using count data from surveys carried out in food premises. In both cases, estimates are greatly dominated by an important source of variability. This variability is related to both analytical method error and experimental variation, as a consequence of there are unknown factors governing bacterial transfer that are not controlled during experimentation (pressure, contact time, attachment, exopolysaccharide production, etc.). In this respect, several studies have considered the use of probability distribution to account for the variability inherent to the bacterial transfer phenomenon (den Aantrekker et al. 2003b; van Asselt et al. 2008; Montville et al. 2001). To give more reliable and complete estimations, probability distributions should be applied to represent bacterial transfers rates between surfaces (Pérez-Rodríguez et al. 2008).

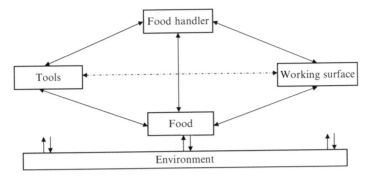

Fig. 4.1 Schematic representation of a bacterial transfer network accounting for bacterial transfer in food-related environments

Fig. 4.2 A simplified bacterial transfer pathway based on transfer events happening successively. *RTE*, Ready to Eat food

The normal distribution is mostly used to fit log-transformed transfer rates (Chen et al. 2001; Montville et al. 2001; Montville and Schaffner 2003; Pérez-Rodríguez et al. 2006), although the beta and gamma distribution have been also successfully used to represent transfer rates (van Asselt et al. 2008; Pérez-Rodríguez et al. 2011).

As already mentioned, one of the most important difficulties associated with modeling transfer is that transfer events can take place in different directions, involving different transfer vectors and leading to a complex bacterial transfer network (Fig. 4.1). In spite of this complexity, some attempts have been made to mathematically reflect transfer pathways related to food handling and manufacturing practices in milk, seafood, and vegetable premises (Aziza et al. 2006; Mylius et al. 2007; Pérez-Rodríguez et al. 2011).

As proposed by several authors (den Aantrekker et al. 2003a; van Asselt et al. 2008; Pérez-Rodríguez et al. 2008, 2006), to cope with such complexity, a systems approach is usually taken in which the network is simplified to obtain specific linear pathways representing the most important cross-contamination routes (Fig. 4.2).

These bacterial transfer pathways can be built by combining individual transfer events (e.g., transfer from food to hand, transfer from hand to food, etc.), which are mathematically described by multiplying the individual transfer rates for each step provided there is independence between transfer events (3.55). As already mentioned, in many cases, such transfer rates are described by probability distributions to reflect the observed ample transfer variability. Therefore, Monte Carlo methods should be applied to perform the transfer calculations.

$$\mathrm{Tr_T}(\%) = \mathrm{Tr_1}(\%) \cdot \mathrm{Tr_2}(\%) \qquad (4.5)$$

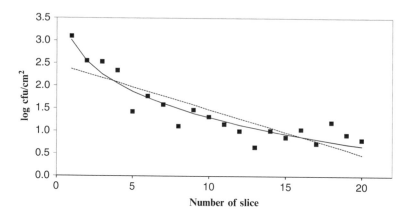

Fig. 4.3 *Escherichia coli* O157:H7 transfer from the contaminated slicer (6 log cfu) to cooked ham during a slicing process. *Solid and dotted lines* represent the Weibull model and log-linear model fitted to transfer data (Adapted from Pérez-Rodríguez et al. 2007a)

During the past few years, more specific transfer models have been built to represent specific bacterial transfer scenarios. There exists a particular interest in explaining bacterial transfer during the slicing and cutting process because many studies have indicated this process is an important contamination source. These studies have intended to describe the bacterial transfer during the slicing process from an inoculated slicer or blade to the product and also from an inoculated food product to the slicer or blade, considering as variable the number of slices (Fig. 4.3). Such studies have used different bacterial species (*Escherichia coli, Staphylococcus aureus,* or *Listeria monocytogenes*) and types of food matrices such as cooked meat, cheese, and salmon, evidencing the influence of the type of microorganism, food matrix, inoculum level, and exerted pressure during slicing on the transfer ability of bacteria (Vorst et al. 2006; Pérez-Rodríguez et al. 2007a; Keskinen et al. 2008; Sheen 2008; Sheen and Hwang 2010). The results have found that bacteria are transferred during slicing following a logarithmic process, and hence, exponential functions have been successfully applied to describe bacterial transfer observations (Sheen and Hwang 2010). For instance, the log-linear and Weibull functions were found by Pérez-Rodríguez et al. (2007b) to be suitable models to describe *E. coli* O157:H7 and *S. aureus* transfer from a contaminated slicer to a cooked meat product during slicing (Fig. 4.3).

Transfer models are mostly based on an empirical approach because we still possess scant knowledge about the mechanisms controlling bacterial transfer in foods. Nonetheless, some attempts have been made to gain insight into the phenomenon, providing some theoretical basis to model transfer in specific situations. In this respect, a probabilistic standpoint has been provided to explain the stochastic nature of the transfer phenomenon of cells based on the binomial process in which observed transfer rates can be understood as the probability of one cell being transferred from one surface to other (Aziza et al. 2006; Pérez-Rodríguez et al. 2008).

More recently, Møller et al. (2012) also proposed a more mechanistic transfer model to explain *Salmonella* transfer in a grinder based on a hypothesis that transfer occurs from two different environments in the grinder showing different transfer rates.

Transfer models are necessary for incorporation in quantitative risk assessment models to yield more accuracy estimations and to contemplate diverse risk scenarios concerning cross-contamination (Pérez-Rodríguez et al. 2008). Hence, this type of model should be considered together with other processes such as bacterial die-off on donor surfaces and food handler behavior to give more accurate estimations of bacterial transfer to food. In this respect, some work has aimed to provide a mathematical framework linking survival models and food handler behavior models to transfer rates models and considering them important variables affecting the final bacterial transfer to foods (Christensen et al. 2005; Ivanek et al. 2004; Mylius et al. 2007).

The bacterial transfer models are still in an early stage because there are many gaps that should be filled in regarding how environmental and intrinsic factors influence such a phenomenon. In spite of this fact, models developed so far seem to well represent the observed transfer rates on an empirical basis, enabling application to quantitative risk assessment studies. Further research together with new modeling and experimental approaches is needed to advance in this new and promising area within predictive microbiology with serious repercussions in public health. These advances will help to derive more reliable mathematical models considering different microorganisms, such as viruses and bacteria, and diverse transfer events.

4.2 Growth/No Growth Models

In the earlier sections, it was stated that predictive microbiology focuses on determining the behavior of a given microorganism, combining mathematical modeling with experimental data under some environmental factors. Predictive microbiology can be divided into two kinds of models in this respect (Tienungoon et al. 2000): kinetic models and boundary (i.e., growth/no growth) models. Boundary models (so-called growth/no growth models) focus their attention on those conditions in which a given microorganism can or cannot grow (Ratkowsky and Ross 1995; Presser et al. 1998; Stewart et al. 2001): the reason is that microbial growth is restricted to finite limits of factors, and even growth sometimes declines abruptly at a very small increase in the level of each factor. Consequently, the combination of both kinetic and probabilistic models can help to provide a complete response of the behavior of the microorganisms in the growth and no growth regions.

Microbial safety should be achieved by maintaining the food at specific conditions that inhibit growth (Salter et al. 2000). To gain knowledge about these conditions, the probability of growth of the microorganism at the boundary zone is needed.

At the same time, food safety has to be attained, avoiding the destruction of the organoleptical characteristics of the food.

Thus, growth/no growth models have been widely used for designing formulations in minimally processed foods, by taking into account the *hurdle technology* concept (Leistner 1992). This is a food preservation technique based on the application of a combination of generally mild treatments that act as 'obstacles' which microflora must overcome to start to grow. Then, bacteria invest their energy in trying to maintain their homeostatic equilibrium instead of multiplying. Although the action mechanisms underlying these treatments are not fully understood, it is very useful to know their effect on bacteria cells as well as the extension of such effects.

Growth/no growth can be used in case of microorganisms for which only their presence can represent a hazard (i.e., spores of *Clostridium* spp.), while kinetic models can be better applied for those nonpathogenic microorganisms or other microorganisms that can be dangerous when exceeding microbial limits, such as *Staphylococcus* spp. (Buchanan 1992).

In this way, probability models are useful in the study of microbial pathogen behavior. The application of these models is clear, for instance, when designing a preservation method in minimally processed foods, for alternative formulations in novel foods, and also for risk assessors so that they can determine the possibility of real contamination of a given food (Presser et al. 1997) or minimize the risk of pathogen growth (McMeekin et al. 2000).

A new trend in the elaboration of growth/no growth models has recently arisen mainly to determine the absolute limits of microbial growth, that is, the presence of extrinsic factors that cannot inhibit growth in themselves but which can when combined (Ross et al. 2000).

Ross and McMeekin (1994) established that growth/no growth models are complementary to kinetic models. However, when maximum growth rate approaches to zero and lag phase to infinite, the microbial behavior should be quantified through growth/no growth conditions. Once a significant growth is produced, predictive microbiology must lead to growth kinetic models.

Ratkowsky and Ross (1995) proposed the application of logistic regression models in food microbiology, which enables the modeling of boundary between growth and no growth of bacteria when one or more controlling factors are studied. This approach included a modified square root model to describe the influence of temperature, pH, a_w, and nitrite on the probability of growth (McMeekin et al. 1993a):

$$\sqrt{\mu_{max}} = a_1 \cdot (T - T_{min}) \cdot \sqrt{(\text{pH} - \text{pH}_{min}) \cdot (a_w - a_{w\,min}) \cdot (NO_{2max} - NO_2)} \quad (4.6)$$

where μ_{max} is maximum growth rate (h^{-1}), T is temperature ($°C$), a_w is water activity, and NO_2 is nitrite concentration (ppm). Minimal values are the theoretical values at which growth is allowed when exceeded. In contrast, maximal values are the theoretical values below which growth is allowed.

By taking the natural logarithm of the left-side term from this equation, one can calculate the term logit P:

$$\text{logit } P = \ln\left(\frac{P}{1-P}\right) \tag{4.7}$$

being P = probability of growth (between 0 and 1, being 0 = no growth and 1 = growth). Thus, the equation shown above is transformed as follows:

$$\text{logit } (P) = b_0 + b_1 \cdot \ln(T - T_{min}) + b_2 \cdot \ln(pH - pH_{min}) + \cdots$$
$$\cdots + b_3 \cdot \ln(a_w - a_{w\,min}) + b_4 \cdot (NO_{2\,max} - NO_2) \tag{4.8}$$

where b_0, b_1, b_2, b_3, y, and b_4 are the regression coefficients to be estimated by the model. The rest of the model parameters (T_{min}, pH_{min}, a_{wmin}, y NO_{2max}) should be estimated independently or are fixed to constant values. This approach was extensively followed by other authors (Lanciotti et al. 2001) and predicts a binary response variable, or equivalently the probability of an event's occurrence in terms of a specific set of explicative variables related to it. Later on, a new nonlinear logistic regression technique was performed (Salter et al. 2000; Tienungoon et al. 2000). Logistic models were further developed to determine growth/no growth interfaces in solid surfaces or food-based systems (McKellar et al. 2002; Koutsoumanis et al. 2004a; Mejlholm et al. 2010).

Growth/no growth models have been implemented to determine the combination of factors that just inhibit or allow growth at a specific probability level. Target values reported in literature for graphical representations are normally set at 0.1 (indicating inhibitory conditions), 0.5 (boundary zone), and 0.9 (high probability of growth). Two illustrative examples are presented in Fig. 4.4 for the models of Kousoumanis et al. (2004b) for *Salmonella typhimurium* and of Valero et al. (2009) for *Staphylococcus aureus* in broth media.

Another growth/no growth model was developed by Masana and Baranyi (2000) to study the growth interface of *Brochothrix thermosphacta* as a function of pH and a_w. The proposed model included a novel parameter named $b_w = \sqrt{1 - a_w}$. Thus, the model shape was divided into two parts: a nonlinear part (corresponding to the relationship between pH and a_w) and a linear one, at a constant value of NaCl.

A cardinal growth/no growth model was performed by Le Marc et al. (2002). This model describes the growth interface of *L. monocytogenes* basing on an original kinetic model with four factors, where μ_{max} is estimated. The final equation is a transformation of the original cardinal model of Rosso et al. (1995).

$$\mu_{max} = \mu_{opt} \cdot \eta(T) \cdot \delta(pH) \cdot \varsigma([RCOOH]) \cdot \phi(T, pH, [RCOOH]) \tag{4.9}$$

where μ_{opt} is the maximum growth rate of the microorganism at optimal conditions.

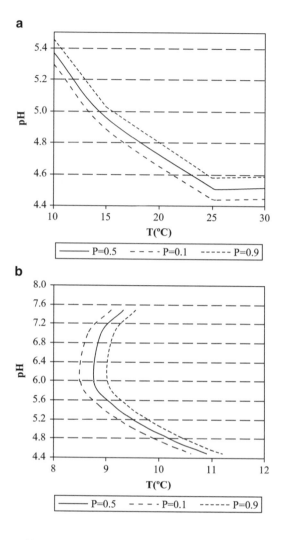

Fig. 4.4 Graphical representation of growth/no growth model predictions for *Salmonella typhimurium* (Koutsoumanis et al. 2004b) (**a**) and for *Staphylococcus aureus* (Valero et al. 2009) (**b**) in broth media at P values of 0.1, 0.5, and 0.9

The growth interface can be easily obtained by reducing the value of μ_{max} to zero. This can be achieved when the interaction term ($\phi(T, \mathrm{pH}, [\mathrm{RCOOH}])$) is equal to zero.

Later, Le Marc et al. (2005) suggested modeling microbial growth limits through the combination of the minimum convex polyhedron (MCP) concept and logistic regression. MCP was introduced by Baranyi et al. (1996) and, according to this definition, the interpolation region is the MCP encompassing all the combinations where measurements were made. The combination of these techniques showed that extrapolation problems at growth limits are solved.

Very recently, a simplified growth/no growth model conceptually derived from the gamma model was proposed (Polese et al. 2011). Bacteria growth limits were through a normalization constant (η), which quantifies the product of the cardinal optimal distances for growth probability, assuming η as a species-independent constant. For complex systems, all the relevant environmental factors could be taken into account (Eq. (4.10)):

$$\Pi = \eta \cdot (T - T_{min}) \cdot (pH - pH_{min}) \cdot (a_w - a_{w\ min}) \cdot \prod_i \left(1 - \frac{C_i}{MIC_i}\right)$$

$$P \cong 0, \ \Pi < 0$$
$$P \cong \Pi, \ 0 \leq \Pi \leq 1$$
$$P \cong 1, \ \Pi > 1 \tag{4.10}$$

where P is the growth probability, T_{min}, pH_{min}, and a_{wmin} are the notional minimal values of temperature, pH, and water activity, below which growth is not possible, C_i is the concentration of the i inhibitory substance, and MIC_i is the minimal inhibitory concentration of the ith substance.

The constant η is defined as

$$\eta = \frac{1}{(T_p - T_{min}) \cdot (pH_p - pH_{min}) \cdot (a_{wp} - a_{w\ min})} \tag{4.11}$$

where T_P, pH_P, and a_{wP} are the theoretical (conceptual) values for each factor (when other factors are at their optimum), above which growth probability is not further affected.

As a whole, the developed simplified model produced conservative estimates, with only a limited number of fail-dangerous predictions, suggesting its potential applicability as a first estimate for the development of safe food products.

These growth/no growth models generally predict an abrupt transition zone between growth and no growth conditions: this occurs because of the small number of replicates (n) in comparison to the number of conditions tested. In this situation, difficulties in the application of logistic regression models are found, especially when achieving convergence to a global optimum or to an appropriate set of conditions (Ratkowsky 2002), unless alternatives procedures are used (Geeraerd et al. 2004).

By increasing the number of replicates ($n \sim 20$–30), smoother transitions between growth and no growth zones can be achieved. With this methodology, probability of growth does not dramatically change along a narrow set of conditions because more reliable growth/no growth responses can be provided (Vermeulen et al. 2007; Valero et al. 2009).

In addition, there is a growing tendency to use a cocktail of different strains to predict a more realistic microbial behavior in foods. In the case of growth/no growth models, as stated by Vermeulen et al. (2007), the use of different strains allows obtaining a broader growth/no growth domain, because at certain stressful conditions, growth is generally led by the most resistant strain. Besides, the importance of taking into account the strain variability for predictive modeling and risk assessment purposes has been emphasized (Lindqvist 2006).

It can be concluded that these mathematical models may lead to more realistic estimations of food safety risks and can provide useful quantitative data for the development of processes that allow production of safer food products (Koutsoumanis et al. 2005).

4.3   Between-Species Interaction Models

Food matrices are complex systems where different microbial populations can coexist and interact. Many studies have demonstrated that microbial flora in foods can inhibit or reduce growth of pathogenic bacteria (Buchanan 1999; Ongeng et al. 2007; Le Marc et al. 2009). However, most predictive models ignore this aspect (Malakar et al. 2003), being able to lead a significant discrepancy between predictions and reality (Vereecken et al. 2003).

Interactions between bacterial populations can be classified into indirect interactions, that is, derived from changes in pH, redox potential, substrates (e.g., antimicrobial substances), or specific metabolites produced by a specific population and direct interaction in which bacterial populations compete for colonization spaces (i.e., niche) and nutrients. All these aspects should be considered in models to better represent microbial response in foods.

Although there are several studies dealing with interaction and competition between bacterial species (Peterson et al. 1964; Carpenter and Chassaing 2004; Jablasone et al. 2005), few have attempted to propose a mathematical basis for explaining such a phenomenon. Leroi and De Vuyst (2007), based on the work by Bernaerts et al. (2004), proposed a general conceptual model explaining interaction between different populations by including the growth-influencing factors.

$$\frac{dN(t)_i}{dt} = \mu_i N_i(t) \tag{4.12}$$

$$\mu_i = f\left\{ \langle N_i(t) \rangle, \langle N_j(t) \rangle_{j \neq i}, \langle E(t) \rangle, \langle M(t) \rangle, \langle S(t) \rangle, \langle H(t) \rangle.. \right\} \tag{4.13}$$

where μ_i is the grow rate of a specific population $i = 1,...n$, $N(t)_i$ and $N(t)_j$ are concentrations of different microbial populations, t corresponds to time, $E(t)$ represents environmental factors, $M(t)$ is microbial metabolite concentration, $S(t)$ is substrate concentration, and $H(t)$ is the physiological state of the concentration of the cells.

Another approach is based on focusing specifically on the inhibitory factors: this is intended to explain the inhibition of a specific population caused by an inhibitory factor derived from other microbial population. As mentioned by Leroi and De Vuyst (2007), this effect can be incorporated into a secondary model by including a new term corresponding to the inhibitory function, denoted by γ, ranging from 0 to

1. This new term is a reduction ratio of the optimum value of the growth rate (μ_{opt}), which can be mathematically defined as

$$\gamma = \mu_i / \mu_{opt} \tag{4.14}$$

here, γ denotes the reduction ratio resulting from an inhibition factor, μ_{opt} is the optimum value of growth rate when the inhibitory factor is absent, and μ_i is the value of growth rate when the inhibitory effect is present.

The inhibition factor γ can be quantified by a mathematical function based on the environmental variables, nutrient depletion, or inhibitory metabolites exerting the inhibitory effect (Leroi and De Vuyst 2007). For example, the γ explaining the inhibitory effect of the lactic acid produced by lactic acid bacteria on *S. aureus* can be defined as follows (Ross and Dalgaard 2004):

$$\gamma = 1 - \left(\frac{[\text{lac}]}{\text{MIC}^0_{\text{lac}}} \right) \tag{4.15}$$

here, [lac] stands for lactic acid concentrations and $\text{MIC}^0_{\text{lac}}$ is the absolute minimal inhibitory concentration of lactic acid.

Then, this inhibitory function is included into a secondary model, for example, based on a cardinal-type model or gamma-concept approach (Leroi et al. 2012). Here, the hypothesis is that the resultant growth of one specific microbial population is caused by the combination of all inhibitory effects derived from both the environmental factors (e.g., pH, temperature, water activity) and the activity of other microbial populations (e.g., acidification, antimicrobial substances). Because independence between effects is assumed, growth can be estimated by multiplying the corresponding inhibitory functions. If no independence is found, more complex terms should be included in the model to explain such an interaction between the inhibition factors (te Giffel and Zwietering 1999).

Mechanistic models have been also proposed to reflect interactions between different microbial populations (Larsen et al. 2012). These (semi)mechanistic models encompasses the multiple factors exerting the inhibitory effect between microbial populations such as nutrient depletion (e.g., nutrient diffusion coefficient), colonization space (e.g., modeling colony growth), production rate of inhibitory or toxic substances (e.g., lactic acid), changes in the chemical properties of food (e.g., pH, water activity), etc. (Malakar et al. 2003; Martens et al. 1999; Poschet et al. 2005). However, as pointed out by Leroi and De Vuyst (2007), the mechanistic approach can lead to complex mathematical models with multiple interrelated variables. Thus, a more simplistic approach is preferred in which microbial interaction modeling is simplified by quantifying how much growth of one population is reduced by growth of other populations. In relationship to this, two different modeling approaches can be taken: one based on the Jameson effect and other based on the Lotka–Volterra type model (Cornu et al. 2011). The former approach, based on the Jameson effect, claims that the interaction between different bacterial populations consists of the growth of lower-density populations slowing down when the higher-density populations reach a maximum level (Jameson 1962). Both approaches can be explained on the basis of the following general growth function:

$$\frac{dN(t)_i}{dt} = \mu_{\max_i} N_i(t) f_i(t) \alpha_i(t) \tag{4.16}$$

where $f(t)_i$ represents an adaptation function and $\alpha(t)_i$ is an inhibition function for a population i.

Cornu et al. (2002, 2011) incorporated the Jameson effect hypothesis into a growth model, assuming that the inhibition effect is equal for all competitive microbial populations, thus $\alpha(t)_I$ is an inhibition function so that $f(t)_i = f(t)_1 = f(t)_2 = f(t)_n$, being a modification of the logistic function of deceleration, is based on the densities of both populations and written as follows:

$$f_i(t) = 1 - \frac{N_i(t) - N_j(t)}{N_{\max_{\text{total}}}} \tag{4.17}$$

This inhibition function has been modified by other authors to explain the effect of specific bacterial populations in determined food matrices (Le Marc et al. 2009; Mejlholm and Dalgaard 2007).

The interaction models based on the Lotka–Volterra type models can be considered as a variation of the previous mathematical structure in which there exist as many functions $f(t)$ as potential competitive microbial populations in which each population has a different inhibition effect represented by the coefficient α_{ij} in the following general equation (Vereecken et al. 2000; Vereecken and Van Impe 2002; Powell 2004):

$$f_i(t) = \frac{1}{N_{\max_i}} \left(N_{\max_i} - N_i - \alpha_{ij} N_j \right) \tag{4.18}$$

where α_{ij} is the interaction coefficient quantifying the effect of one population (i) on the other (j).

Including the effect of competition between species into predictive models is a needed step to achieve more accurate predictions. Indeed, the recent application of bio-protective cultures as food preservatives and the need to know how endogenous flora affects growth of pathogens in foods is increasing the attention of microbiologists to this area as models considering these factors might have important applications in the food industry. Although between-species interaction models are still incipient, the efforts made until now have resulted in suitable mathematical models focused on the paramount inhibitory effect produced by the dominant microbial population. Future trends in modeling the interaction between bacteria species move toward a mechanistic approach considering bacterial growth in foods from a more holistic perspective (Larsen et al. 2012).

4.4   Single-Cell Models

Modeling the variability associated with the lag phase of microorganisms constitutes a significant aspect of a microbial risk assessment. Previous research in this field has widely demonstrated that variability sources are attributed to the lag phase rather than to maximum growth rate (McKellar et al. 1997; Augustin and Carlier 2000a). As already mentioned, growth of microorganisms in broth media can underestimate lag phase because of ideal existing conditions. Similarly, other additional factors such as inoculum level or pre-incubation conditions (cell history) influence lag-phase values. As microbial contamination of foods normally occurs at low levels, higher variability in lag phase is expected, especially when cells are stressed (Robinson et al. 2001). In contrast, higher population levels increase the probability of finding at least one cell to start multiplication, originating shorter lag-phase values (Baranyi 1998). Augustin et al. (2000) showed that the lag phase was longer when cells were severely stressed and the inoculum level was lowered. This inoculum-level effect can be explained by an increasing variability in individual cell lag phases when stress factors become more stringent. Several authors have already studied the effect of environmental conditions on the distributions of the individual cell lag (Smelt et al. 2002; Métris et al. 2003, 2008; Francois et al. 2006). These observations were confirmed by Métris et al. (2002) when the effect of salt stress and acid stress on lag-time distributions of isolated *L. monocytogenes* cells was investigated.

Moreover, studies at the single-cell level are needed to better quantify this variability and predict microbial behavior in realistic conditions. The individual-based approach of the lag phase is gaining interest, especially for pathogens that initially contaminate food products in low amounts. In turn, this could be a major drawback in predictive microbiology, because counting methods can only measure cell concentrations higher than 10 cfu per sample. Inaccuracies in measurements can cause difficulties in model validation. Growth modeling concentrates on the early stages of microbial growth in the actual environment. The previous environment (i.e., pre-incubation conditions) can largely affect growth in the actual environment, whose effect gradually decreases after inoculation: this is called the *adjustment process*, as described by Baranyi (1998). After adjustment, an exponentially growing population is established that ultimately reaches an upper limit (stationary phase). This population limit is caused by the depletion of nutrients or by the accumulation of waste products of metabolism.

Modeling at the single-cell level uses stochastic processes when variability among individual cells is taken into account. An intrinsic feature of stochastic processes is that one can obtain unexpected results. Relationships between the individual lag-time distributions and the lag time of the bacterial population show, for instance, that the doubling time depends on the number of cells, from $1/\mu_{max}$, when the number of cells is 1, converging to $\ln2/\mu_{max}$, at higher concentration levels (Baranyi and Pin 2001).

The lag phase and subsequent growth before the stationary phase can be modeled using a biphasic linear function:

$$\ln(N_t) = \ln(n) + \mu_{max}(t - L) \tag{4.19}$$

where N_t is the population size at time t, n is the inoculum size, μ is the specific growth rate, and L is the population lag phase.

Malakar and Barker (2008) stated that individual lag phases of individual cells in the inoculum, L_i, are identically and independently distributed random variables. Thus, the natural logarithm of the population with a sufficient long time is (when t is higher than L_i)

$$\ln(N_t) = \ln(n) + \mu \left[t - \frac{1}{\mu} \left(\ln \frac{\sum\limits_{i=1}^{n} e^{-\mu \cdot L_i}}{n} \right) \right] \tag{4.20}$$

Then, from the biphasic growth model, the population lag time K_n, arising from an initial inoculum of size n, is

$$K_n = -\frac{1}{\mu} \left(\ln \frac{\sum\limits_{i=1}^{n} e^{-\mu \cdot L_i}}{n} \right) \tag{4.21}$$

This formula shows that lag phase decreases with the initial cell number because the exponentially growing subgenerations of cells with shorter lag phases will take over the whole population.

The three-phase linear model proposed by Buchanan et al. (1997) has a physiological basis. It is assumed that each individual cell has a certain lag phase, L_i, and a certain generation time, t_m. Furthermore, the lag-phase duration of a single cell is subdivided into two parts, namely, the adjustment period and the metabolic period. During the first period, the cells adapt themselves to their new environment. The metabolic period is the time needed for the cell to generate sufficient energy and synthesize biological materials needed for cell replication.

To simulate transition between nongrowing and growing cells McKellar and Knight (2000) developed a discrete–continuous model. This model combines a discrete adaptation step, as a property of individual cells, with the continuous logistic model for bacterial growth. An extension to a continuous–discrete–continuous model included dynamic conditions for modeling population lag because μ_{max} can change as function of environmental conditions during the lag phase.

Baranyi and Pin (2001) described the mathematical relationship between the individual and population lag phase. They concluded that distributions of individual lag phases are different from the distributions of the population lag phases.

However, at the current accuracy of available data, it is impossible to deduce the distribution of the individual lag phases from a population growth curve. Moreover, to study the distribution of lag times of individual cells, a large quantity of replicate measurements is necessary. Consequently, the lag phase of individual cells cannot be studied using traditional viable counts.

Automated turbidity measurements could provide a solution because they are suited to producing a large quantity of replicate detection time measurements. If each observed culture starts from a single cell, then the distribution of the detection times should be close to the distribution of the lag times of the initial individual cells, assuming that the specific growth rate is the same and constant for each population engendered by a single cell (Métris et al. 2002, 2003).

In this sense, protocols to isolate single cells have been optimized by Francois et al. (2003) for *Lactococcus lactis*, and subsequently developed for *Listeria monocytogenes* (Francois et al. 2006), through the use of turbidimetric measurements in microtiter plates. This method combined a high probability of having single cells with a sufficient yield. A similar approach was followed by Guillier et al. (2005) by assuming a Poisson distribution of the inoculated cells in the microtiter wells to obtain single cells. Then, the growth of the population generated by that cell can be detected and monitored by an automatic turbidimeter, such as the Bioscreen (Labsystems, Finland). The time elapsed until the detection of growth is used to derive the lag time for the original single cells.

Wu et al. (2000) compared the dilution method with a microscopy method for determination of the lag time of *L. monocytogenes* and concluded that microscopy provides accurate estimates of the lag time of single cells. The use of microscopy has several advantages compared to the dilution method for the determination of the lag phase of single cells. First, the microscopy method is a direct method, allowing direct observation of the first cell division, whereby the other is indirect; it is based on the (turbidity) detection time of the one-cell-generated subpopulation. The calculation of the lag phase from detection times is based on assumptions that are difficult to check. Also, the variation of the subsequent generation times also contributes to the variation of the observed detection times. The need for direct methods to study the lag phases of single cells was also stressed by McKellar and Knight (2000). If a treatment results in cells not dividing, the detection time-based methods can establish only 'ND' results ('not detected' within the observation time), whereas microscopy-based methods can distinguish between dead and live cells (Wu et al. 2000). These methods were used by other authors to evaluate the effect of preservatives on *Listeria innocua* in solid agar surfaces (Rasch et al. 2007). Similarly, Guillier et al. (2006) introduced a new method of the estimation of individual lag phases based on the measurement of bacterial colony surfaces by an image analysis procedure.

Wu et al. (2000) observed that the difference between lag times obtained from microscopy data and lag times computed from (DT) data is that the former observes when the cell effectively divides for the first time (t_{lag}) and the latter observes when the cell biomass starts to grow, that is, increase in volume(s). Both lag values were related by the equation:

$$Li = \tau + Dob_t \qquad\qquad (4.22)$$

where L_i is the mean individual cell lag time, τ the adaptation time, and Dob_t the doubling time.

Stochastic approaches to study variability of individual cells at growth boundaries were developed by Koutsoumanis (2008). He described that the non-growing fraction of cells close to the growth boundary resulted in a delay of population growth, which he called pseudo-lag. The environment determines the extent of pseudo-lag whereas its variability is affected by both the inoculum size and the growth conditions. At growth-limiting conditions the total *apparent lag* of the population is a convolution of the *pseudo-lag* and the *physiological lag* of the growing cells. These studies are becoming more interesting at conditions approaching growth limits, where the variability of cell behavior increases.

Chapter 5
Software and Data Bases: Use and Application

Abstract Obtaining data for improving food safety management systems is often required to assist decision making in a short timeframe, potentially allowing decisions to be made and practices to be implemented in real time. Collection, storage, and retrieval of new data regarding microbial responses in foods gain insight on the achievement of food safety management measures (i.e., food safety objectives, performance objectives), avoiding the increase of fail-dangerous events. The role of data bases in predictive microbiology has been widely demonstrated as useful tools for the development of computing software or tertiary models, which allow users to estimate growth, survival, or inactivation of food-borne pathogens and spoilage microorganisms in different food matrices. Additionally, the fast development of information and communication technologies (ICTs) has increased the software tools available in predictive microbiology. These tools, named tertiary models, are created for a wide range of applications and types of users: scientists, food operators, risk managers, etc. Although early versions were designed as standalone systems, nowadays on-line software is a major trend making tools available everywhere to everyone through the Internet. In this chapter, descriptive examples of data bases and software tools used in predictive microbiology are explained.

Keywords Data base • ComBase • Sym'Previus • Microbial Response Viewer • Pathogen Modeling Program • ComBase Predictor • Seafood Spoilage and Safety Predictor • MicroHibro

5.1 The Data Base as a Source of Data for Modeling Purposes

As stated in the earlier sections of this brief, one of the main objectives pursued in predictive microbiology is quantification of microbial responses in foods. However, the complexity of the food environment has been recognized, which thus makes it difficult to quantify or even to categorize some of its features and their potential

F. Pérez-Rodríguez and A. Valero, *Predictive Microbiology in Foods*,
SpringerBriefs in Food, Health, and Nutrition 5, DOI 10.1007/978-1-4614-5520-2_5,
© Fernando Pérez-Rodríguez and Antonio Valero 2013

effects on microbial population dynamics or the ability to recover a target organism from a food sample. An additional difficulty is that, with the background information on the environment and with currently available techniques to measure microbial responses, both variability and uncertainty may be large (Ratkowsky 2004), which can produce wider confidence limits and a reduction in the accuracy of microbial estimates. It should be noted that predictive models themselves constitute a simplification of the real biological responses of microorganisms affected by different combinations of physicochemical and biological environmental factors. When these factors are identified, the same simplification is carried out as when a process is characterized by some mathematical variables.

Predictive microbiology software programs are based on data bases and mathematical models. Behind predictive software programs are the raw data upon which the models are built. The relationship between mathematical models and data bases provides the fast comparison of large amounts of data under a standardized and harmonized recording format.

A data base is a large collection of data organized in a specific form for rapid search and retrieval. Creation of data bases in predictive microbiology can provide a solution to inaccuracy in measurements, which can be compensated by increasing their number, to variability in responses and to data exchange between different research and academic institutions.

The largest and most used database, called *ComBase* (Combined, or Common, i.e., joint, Database of microbial responses to food environments) was launched at the Fourth International Conference on Predictive Modeling in Foods, Quimper, France, June 2003. Its technical details can be read in Baranyi and Tamplin (2004) and on the website (www.combase.cc).

This data base was developed in the Institute of Food Research, Norwich, UK (IFR), to pool available predictive microbiology data. Soon, the leaders of FSA and USDA-ARS agreed that incorporating all their data in this common data base would be mutually beneficial. The European Commission also embraced the idea, and now the original Food MicroModel and PMP datasets have been supplemented with additional data submitted by supporting institutes, universities, and companies, mainly from Europe.

Furthermore, data have also been compiled from the scientific literature and a continuous updating process is carried out. The website is maintained by the IFR in collaboration with the Food Safety Center, University of Tasmania, and the Eastern Regional Research Center of the USDA Agricultural Research Service.

Accessibility and application of resources of data involves thousands of researchers, risk assessors, legislative officers, food manufacturers, and their laboratory managers at no expense.

Users can compare observations with independent predictions gained from other software packages or with external data. If *ComBase* is accepted internationally as the benchmark, the number of sources generating different views on risk can be decreased (McMeekin et al. 2006).

In Fig. 5.1 can be seen a explanatory record in *ComBase*. The raw data are organized in the data base and a browser is incorporated to facilitate data searching.

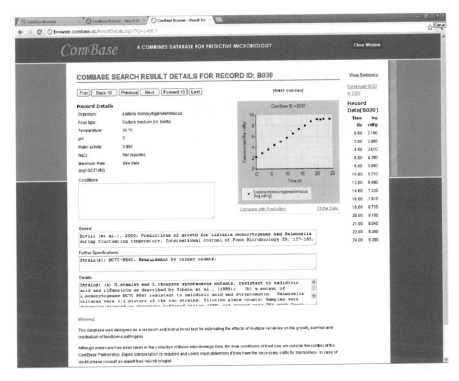

Fig. 5.1 Example of a detailed record in *ComBase*

Additional information about physicochemical factors, culture, media, and material and methods followed are provided. Observed data (graphical representation and numeric data) can be used for academic or scientific purposes.

Although *ComBase* contains a vast amount of data (more than 50,000 records), sometimes it is not easy to obtain the desired information from the retrieved data, especially the case of a growth/no growth interface, where variability is much larger. However, Le Marc et al. (2005) showed an application of the use of *ComBase* data within the MCP concept to generate growth/no growth models for various pathogens.

For a better understanding of Combase information, a new ComBase-derived data base was developed by Koseki (2009), named Microbial Responses Viewer (MRV). The software can be accessed at http://mrv.nfri.affrc.go.jp/Default.aspx#/About. A screen capture is presented in Fig. 5.2. The main objective is that food processors or other interested partners can easily find the appropriate food design and processing conditions from the retrieval of microbial growth/no growth data. The response was defined as representing 'growth' if a significant increase in bacterial concentration (>1.0 log) was observed. Alternatively, 'growth' was defined as a positive value of the specific growth rate. The specific growth rate is illustrated using a two-dimensional contour plot with growth/no growth data.

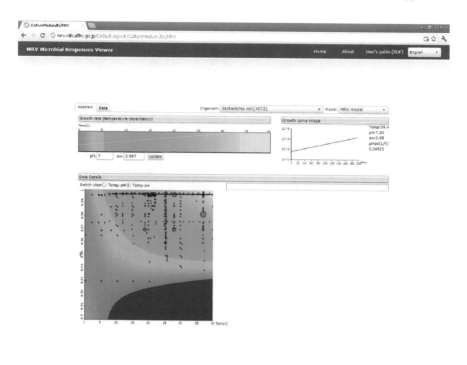

Fig. 5.2 Microbial Responses Viewer. Overview of the growth model of *Escherichia coli* in culture media (*upper part*) and a contour plot (*bottom part*) of growth/no growth interface together with a comparison with observed data retrieved from *ComBase*

MRV provides information concerning growth/no growth boundary conditions and the specific growth rates of queried microorganisms.

The software allows the user to rapidly view growth/no growth contour plots superimposed by actual ComBase data. Contours of any two of three variables (temperature, pH, and a_w) can be visualized, while the third is held constant.

The Sym'previus project (www.symprevius.org) is an extensive decision support system developed in France that includes a data base and simulation tools for growth, survival, inactivation and growth/no growth interface of pathogenic bacteria and some spoilage microorganisms. Consumer exposure can be evaluated by means of a probabilistic module. Information from Sym'previus is available on a commercial basis through contact centers as indicated on the homepage cited above. Among the software features, a data base was built to integrate food, bacteria, and environmental characteristics on microbial behavior concerning pathogenic germs able to contaminate food, and also epidemiological data (prevalence or level of contamination in food).

The Sym'Previus network started in 1999. Sym'Previus meets the French expertise in predictive microbiology of major food companies, technical centers, and public research institutes. This project aimed at proposing an assisting tool in the management of food safety. The data base is integrated into a query system, called MIEL, a multi-criterion system that allows formulating specific interrogation with a specific selection of food and microorganisms. MIEL crosses the user demand about food, microorganisms, and environmental factors (Leporq et al. 2005). Therefore, one can obtain the data most closely related to the searched information.

In a further step, a number of softwares called expert systems have been developed to provide more complex decision support based on a set of rules and algorithms for inference based on the relationships underlying these rules. In these systems, the core knowledge is stored as a series of IF–THEN rules that connect diverse evidence such as user input, data from data bases, and the formalized opinions of experts into a web of knowledge. There are various examples of decision support systems that encapsulate knowledge which is based in predictive microbiology, such as those applied in predicting food safety and shelf life (Zwietering et al. 1992); a stepwise system structured as a standard risk assessment process to assist in decisions regarding microbiological food safety (van Gerwen et al. 2000); or other systems for microbial processes as described by Schellekens et al. (1994).

5.2   Fitting Software for Modeling Purposes

Nowadays, the great development of computational sciences and software engineering have enabled obtaining user-friendly software that, based on specific mathematical algorithms, are able to fit many types of mathematical function to observations including qualitative and quantitative data. In predictive microbiology, observations usually correspond to either counts describing microbial growth and inactivation or kinetic parameter values obtained under different environmental conditions. In such cases, different fitting procedures can be used depending on the type of model and variable considered. As mentioned in a previous section, linear and nonlinear regression methods are applied to data on the basis of the type of mathematical function to be fitted. Most software incorporates both types of fitting procedure and, in some cases, specific developments and algorithms are implanted to optimize fitting procedure.

The MS Excel adding DMfit is a free software application for predictive microbiology modeling developed by the Computational Biology group at Institute of Food Research (Norwich, UK). The application is a Microsoft Excel Add-In to fit log counts versus time data, providing kinetic parameters such as growth/death rate and lag time/shoulder. The available primary models are the reparameterized Gompertz equation (Zwietering et al. 1990) and the Baranyi model (Baranyi et al. 1995), including a modification for fitting optical density data. Furthermore, the application includes a module to fit secondary models encompassing the gamma model, Ratkwosky model, and polynomial models. The latest version of Dmfit can be downloaded at the Combase website http://www.combase.cc/. In addition, there is an on-line version, which can be executed in the Internet, including some basic features taken from the Excel macro. This version is also available through the combase's webpage.

GInaFiT is a freeware add-in for Microsoft Excel aimed to fit several inactivation models (Geeraerd et al. 2005) to experimental data provided by users in an Excel spreadsheet. This application has been developed by the chemical and biochemical process technology and control group (Biotec) at the University of Leuven. The application includes a collection of the most representative inactivation models highlighting the log-linear model (e.g., Bigelow model), and the Weibull model and its adaptations. The software can be downloaded on-line at http://cit.kuleuven.be/biotec/downloads.php and once installed, executed in the Microsoft environment. Similar to this application, an Excel macro developed by Prof. M. Peleg of the Department of Food Science at University of Massachusetts enables fitting the Weibull model to inactivation data. The macro files can be downloaded at a specific website containing other similar macros concerning predictive microbiology developed by Prof. Peleg's research group (http://people.umass.edu/~aew2000/GrowthAndSurvival/GrowthAndSurvival.html).

Symprevius is a decision-making tool based on predictive microbiology with different features that is discussed in the next section. Among them, we highlight here the growth-fitting model tool. This tool is able to fit primary and secondary models to data provided by users based on the primary model function proposed by Rosso (1995) and cardinal values-type secondary model (Rosso et al. 1995). The application fits models applying a subroutine taken from R software's so-called 'nls' function (Leporq et al. 2005). These tools are available on-line by simply taking out a subscription at http://www.symprevius.org.

More specific subroutines have been developed by using R software. In relation to this, the growth curve-fitting package, so-called 'grofit,' developed by Kahm et al. (2010) is able to fit growth functions such as the logistic function, Gompertz equation, modified Gompertz equation, and Richards equation studied in the work by Zwietering et al. (1990).

More general statistical software can be also used by predictive microbiology users to fit models. The most used software are from SAS (SAS Institute, Cary, NC, USA), Matlab (Mathworks, Natick, MA, USA), SPSS (SPSS, Chicago, IL, USA), or R (R Development Core Team), software that includes a multitude of fitting options to perform linear and nonlinear regression procedures based on the maximum likelihood and least squares methods. For other more accessible alternatives, users who are not familiar with complex statistical packages based on a specific program language may turn to more user-friendly applications such as GraphPad Prism software (GraphPad, San Diego, CA, USA) or TableCurve software (SYSTAT Software, Richmond, CA, USA), which also allow fitting linear and nonlinear functions.

Microsoft Excel (Microsoft, Redmond, WA, USA) is widely used by scientists from different research areas. Predictive microbiology is not an exception and, indeed, some curve-fitting tools are built in Excel, as already mentioned. Excel can also be used to fit curves in a simple way by using the Solver function, which enables one to find an optimal solution based on an iterative process and the constraints defined by users. Linear and nonlinear regression procedures can be applied in Excel with the least squares method or the maximum likelihood method (Brown 2001). The optimal solution, that is, the values of regression parameters that best fit the observations, can be found by combining search algorithms (e.g., Newton and conjugate methods) and algorithms for providing initial estimates of variables (e.g., tangent and quadratic methods), which can be chosen according to the type of function to be fitted and available computational resources (Walsh and Diamond 1995). Some examples of Excel applications to fit nonlinear functions have been shown in predictive microbiology literature (Koutsoumanis et al. 2006; Corradini and Peleg 2007). For predictive microbiology beginners who desire to deepen the use of Excel for fitting growth curves, we would recommend reading thoroughly the book by Billo (2007), which presents in an understandable way the main applications of Excel in science.

5.3   Prediction Software: Some Examples

The introduction and rapid spread of information and communications technologies (ICTs) have also reached predictive microbiology. During the past few years, numerous software concerning predictive microbiology (e.g., tertiary models) has been developed, covering a wide range of applications and type of users: scientists, food operators, risk managers, etc. Although early versions were designed as standalone systems, nowadays, on-line software is a major trend, making tools available everywhere to everyone through the Internet.

The Pathogen Modeling Program (PMP) and ComBase Predictor can be considered to be the pioneering software about application of predictive microbiology models. Both applications are based on the same philosophy, to make available unpublished and published models developed by official organisms and scientists by incorporating them into an easy-to-use software tool for end-users. PMP software is developed by the U.S. Department of Agriculture-Agricultural Research Service (USDA-ARS) and particularly at the USDA-ARS Eastern Regional Research Center (ERRC) in Wyndmoor, Pennsylvania. The distribution of the application started in the 1990s and has continued to the present time. PMP is built upon predictive models developed by USDA, which mainly are growth and inactivation (irradiation and pasteurization) models for different pathogenic bacteria (*Listeria monocytogenes*, *Escherichia coli* O157:H7, *Bacillus cereus*, etc.). The primary models in PMP are based on the Gompertz equation (Zwietering et al. 1990). Users are provided with estimates of generation time, lag time, kinetic curve graphs, and their confidence intervals for the selected values of environmental factors (Fig. 5.3). The PMP software, which is now in its seventh version, PMP 7.0, can be downloaded at the PMP website http://www.ars.usda.gov/Services/docs.htm?docid=11550. Recently, PMP has launched an on-line version of PMP software using a selection of the predictive models included in the standalone version in addition to new developed models, among which we can highlight the module dealing with transfer models.

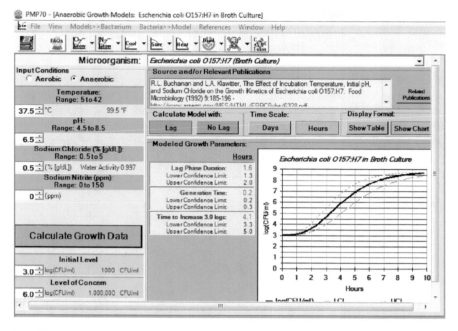

Fig. 5.3 Screen from the Pathogen Modeling Program (PMP)

ComBase Predictor is a free on-line tool, as is PMP online which enables us to predict the response of different types of bacteria to key environmental factors in food (temperature, pH, salt concentration, etc.). ComBase Predictor has been developed at the Institute of Food Research (Nowrich, UK). Also, predictive models were developed by this research group on the basis of kinetic data obtained in culture broth. This application is the successor to the UK Food MicroModel program described by McClure et al. (1994). The primary model used by ComBase Predictor is the Baranyi's model (Baranyi and Roberts 1994), and the secondary models are polynomial equations relating environmental factors and kinetic parameters. In contrast to PMP, Combase Predictor software predicts growth and survival of microorganisms as a function of temperature, pH, and salt concentration, including in some cases the effect of a fourth environmental factor, such as the concentration of carbon dioxide or organic acids (Fig. 5.4). In addition, ComBase Predictor also includes spoilage microorganisms such as certain species of lactic acid bacteria. As an important feature of this software, which makes it different from PMP, is that ComBase Predictor allows predictions under dynamic temperature, permitting introducing time–temperature profiles for all microorganisms considered in the application. Therefore, users are able to introduce data recorded by temperature loggers obtaining growth or inactivation predictions for the introduced profile. Furthermore, ComBase Predictor can simultaneously produce predictions for up to four microorganisms. The application is accessible on-line after registration at the ComBase website (www.combase.cc).

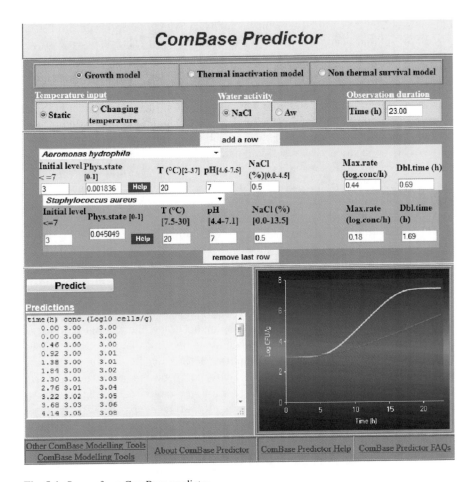

Fig. 5.4 Screen from ComBase predictor

Another important prediction software is the Seafood Spoilage and Safety Predictor (SSSP), which exclusively deals with predictive models for seafood and fish products. The application was developed by the DTU National Institute of Aquatic research. In a first version, the application only contained spoilage predictive models for different types of seafood products (smoked salmon, shrimps, etc.), which was known as the Seafood Spoilage Predictor (SSP). The current format of SSSP, launched in 2004, incorporates a food safety option focused on *Listeria monocytogenes* growth (Gimenez and Dalgaard 2004). The main features of the SSSP software are (1) inclusion of the relative rate of spoilage models for products from temperate and tropical waters based on sensory analysis, (2) spoilage models based on growth of microorganisms such as *Photobacterium phosphoreum* and *Shewanella putrefraciens*, (3) predictions under dynamic conditions of temperature using time–temperature profiles from data loggers and user-defined predictive models based on the cardinal-type model approach (Mejlholm and Dalgaard 2007). The SSSP software can be downloaded at the website of DTU National Institute of Aquatic Research (http://sssp.dtuaqua.dk). The latest version corresponds to SSSP v. 3.1, which is translated into several languages including English, Spanish, and Chinese.

Symprevius is another software tool available on-line developed by INRA (French National Institute of Agricultural Research) and other public and private partners that includes several predictive microbiology functionalities (Leporq et al. 2005). This application requires a commercial subscription for full access. Basically, Symprevius allows making predictions concerning growth, inactivation, and growth/no-growth interfaces for different microorganisms and foods under statistics and dynamic conditions. Similarly, it permits including a stochastic component in variables to represent variability and uncertainty in estimates. In addition, specific tools are specially designed for strengthening HACCP systems and developing new products or determining shelf life. The application can be accessed at www.symprevius.net.

As already mentioned, predictive models are equations or mathematical functions that are derived from experimental data to provide estimates of microbial response in real systems. However, it is well known that predictive models and particularly complex mathematical functions are not accessible for nonexpert users. Hence, when models are built, an applicability component should be incorporated making the model ready to apply. Computing sciences and software engineering allow us to provide models with an applicability component, converting models not only in suitable function describing observation, but rather tools to be applied by end-users. A new prediction on-line software tool was launched in 2012, developed by the Predictive Microbiology Group at University of Córdoba. The application incorporates different predictive models for a variety of microorganisms and foods, enabling users to obtain growth and inactivation predictions under selected environmental conditions (Fig. 5.5). The application, so-called MicroHibro, claims a new dimension for predictive models concerning its application and usability. MicroHibro allows including any type of mathematical function enabling its easy update and making the tool dynamic and renewable. The on-line application can be freely accessed after user registration at http://www.microhibro.com. Because MicroHibro is an on-line tool, users can save its own predictive models and predictions in a virtual account, which can be accessed anytime and anywhere to retrieve saved data. MicroHibro also incorporate a validation module to allow users to assess available models using their own data. Finally, the application applies a stochastic approach intended for risk assessors to carry out probabilistic risk models based on an object-oriented system and allowing defining environmental factors as probability distributions. Results are displayed using a suitable graphic interface to improve interpretability and data analysis.

The increasing interest in predictive software tools has produced a multitude of specific prediction software tools, focused on determined food categories or industrial sectors. Even private companies such as PURAC have developed prediction software tools for *Listeria* (Opti. Form, Listeria Suppression Model, and Listeria

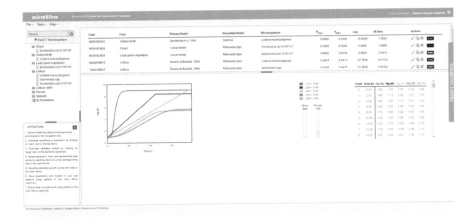

Fig. 5.5 Screen from MicroHibro software

Control Model). In general, PURAC software tools deal with *L. monocytogenes* growth, including, as factors, preservatives commercialized by PURAC and other environmental factors influencing *Listeria* growth. PURAC offers this tool (as adding value to their products) to help users and customers obtain safer product formulations. Some of these applications can be freely downloaded at the PURAC website http://www.purac.com/EN/Food/Purac-Calculators/Listeria-Control-Model.aspx with previous registration. Similarly, Campden and Chorleywood Food Research Association, Chipping Campden, UK, commercializes a software tool for estimating shelf life and formulation of bakery products. More recently, two works by Psomas et al. (2011, 2012) presented a standalone software tool, the so-called UGPM (Unified Growth Prediction Model), based on Visual Basic programming language. The application provides predictions of microbial growth in foods under dynamic or static temperature conditions. In particular, the UGPM software applies the primary model of Baranyi and Roberts (1994), in combination with secondary temperature models, to simulate growth of a spoilage and pathogenic bacteria (*Staphylococcus aureus, L. monocytogenes*, and lactic acid bacteria) during storage of a specific food or food category (e.g., cheese and cooked meat products). UGPM is built upon a specific mathematical algorithm to estimate growth under dynamic conditions on the basis of Baranyi's growth model (Baranyi and Roberts 1994). This software can be freely downloaded at the authors' website: http://users.uoa.gr/~apsomas and http://www.aua.gr/psomas.

Chapter 6
Application of Predictive Models in Quantitative Risk Assessment and Risk Management

Abstract Food-borne pathogens associated with food products are a major concern of both industries and governments; thus, design of proper risk mitigation and elimination strategies is required. Currently, the great development showed by the scientific method and the tendency to optimize processes through its systematization has led to the necessity to unify and standardize food safety management processes. With this, it is not intended to abandon the approach that has prevailed historically, based on consultations of experts and use of 'default values' (conservative control limits and measures which establish the guarantee of the safety of a process or food), but complete foundations to improve its result through a structured approach based on scientific facts. In this respect, the World Trade Organization (WTO) (The WTO agreement on the application of sanitary and phytosanitary measures (SPS Agreement), 1995), according to the agreements of the General Agreement on Tariffs and Trade (GATT) and Sanitary and Phytosanitary Measures (SPS), proposes that to ensure fair and safe international trade, standards and harmonized food regulation need to be established, based on a scientific and rigorous approach, recommending for that the application of methods of risk assessment. Application of predictive models within a risk assessment framework is presented throughout this chapter.

Keywords Risk analysis • Quantitative microbiological risk assessment • Modular process risk model • Food Safety Objective • Performance Objective • Uncertainty • Variability • Probability distributions

6.1 The Risk Analysis Framework

The process of conducting a Microbial Risk Assessment is a structured, systematic approach to integrate and evaluate information from diverse sources concerning the origin and fate of pathogens along the food chain and to determine the magnitude of

F. Pérez-Rodríguez and A. Valero, *Predictive Microbiology in Foods*,
SpringerBriefs in Food, Health, and Nutrition 5, DOI 10.1007/978-1-4614-5520-2_6,
© Fernando Pérez-Rodríguez and Antonio Valero 2013

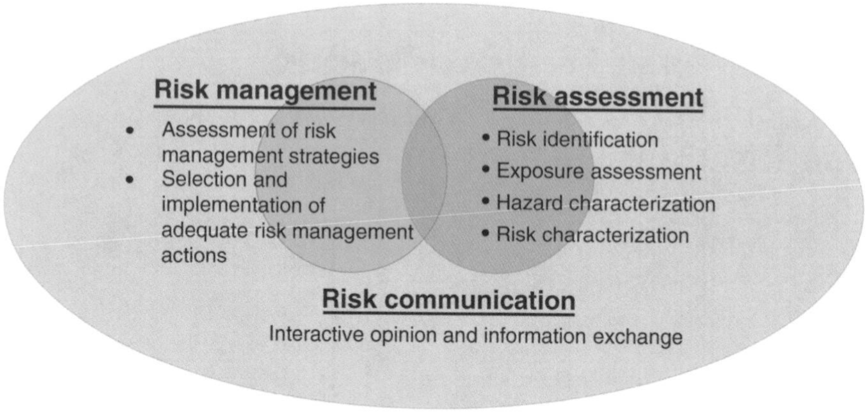

Fig. 6.1 Interaction between the elements of risk analysis (FAO/WHO 2006)

public health risks. In addition, predictive microbiology can help risk assessors and risk managers to make decisions concerning risk mitigation in food products because it is possible to know the microbial behavior in a medium, or in a food through different mathematical models, as a function of certain intrinsic or extrinsic environmental factors.

FAO/WHO (Food Agriculture Organization/World Health Organization) taking the lead from WTO in 1995, introduced, through meetings of experts, the concept of risk analysis as a systematized and 'rational' scheme in which the development of food standards on national and international scale is based (FAO/WHO 1995). This approach also could be followed by food industries and food companies, even though in this case its development is mainly intended to improve the Hazard Analysis and Critical Control Points (HACCP) programs and to assess, from the bio-sanitary point of view, new designs and novel products (van Gerwen and Gorris 2004; Voysey 2000). This food safety management approach has been completed and developed through the inclusion of other concepts such as the Food Safety Objective (ICMSF 2002). The use of predictive modeling will help in choosing the most appropriate levels of factors to be used to meet target management measures. The main achievement within a HACCP system lies on setting quantitative levels, which can be transmitted to governments and national authorities to improve food safety.

The Risk Analysis, according to the FAO/WHO (1995), consists of three components (Fig. 6.1): Risk Assessment, Risk Management, and Risk Communication.

Risk assessment is 'the qualitative and/or quantitative evaluation of the nature of the adverse effects associated with biological, chemical, and physical agents which may be present in food' (FAO/WHO 1995). Risk assessment is structured in four steps: Hazard Identification; Hazard Characterization, Exposure Assessment, and Risk Characterization. The document entitled Principles and Guidelines for the

Conduct of Microbiological Risk Assessment (Alinorm 99/13A) provided specific definitions for the four steps of Microbiological Risk Assessment:

1. Hazard Identification. 'The identification of biological agents capable of causing adverse health effects and which may be present in a particular food or group of foods'.
2. Hazard Characterization. 'The qualitative and/or quantitative evaluation of the nature of the adverse health associated with the hazard'.
3. Exposure Assessment. 'The qualitative and/or quantitative evaluation of the likely intake of a biological agent via food, as well as exposure from other sources if relevant'.
4. Risk Characterization. 'The process of determining the qualitative and/or quantitative estimation, including attendant uncertainties, of the probability of occurrence and severity of known or potential adverse health effects in a given population based on hazard Identification, Hazard Characterization and Exposure Assessment'.

The variety of the necessary knowledge to manage correctly a risk assessment requires a cross-disciplinary team (microbiology, epidemiology, medical science, food technology, etc.) that can handle appropriately the available scientific information.

Risk assessment is the basis of the following component in the risk analysis framework, that is, the risk management. Risk management is defined by the FAO/WHO (1997) as 'the process of weighing policy alternatives in the light of the results of Risk Assessment and, if appropriate, selecting and implementing appropriate control options including regulatory measures.' On this basis, the Codex Alimentarius has developed its own procedure to elaborate Codex Standards (FAO/WHO 2001).

The last component of the risk analysis is risk communication, defined as 'an interactive exchange of information and opinions concerning risk among risk assessors, risk managers, consumers and other interested parties' (FAO/WHO 1998).

These three components should be functionally separated to avoid whatever type of conflict of interest. Nevertheless, it has to be considered that, at the same time, the risk analysis is an interactive process in which the interaction between risk assessor and risk manager, with regard to its practical application, is essential (Fig. 6.1) (FAO/WHO 2006).

The risk analysis process should be evaluated and reviewed as appropriate, with the aim of generating new scientific data. Similarly, the principle of precaution should be followed as many sources of uncertainty exist in the process of risk assessment and risk management of foods related to human health. Thus, the reported results may state the degree of uncertainty assumed in each option and the characteristic of the hazard. More principles are described in detail by the Codex Alimentarius (2007).

Recently, FAO/WHO have published a guideline to support countries in applying risk analysis principles and procedures during emergencies in their own national food control systems, as risk analysis as a key component of national Food Safety Emergencies Response planning (FAO/WHO 2011). It is concluded that the application of risk analysis during an emergency should follow the same principles applied under normal circumstances in each country. The only differences in an emergency situation are the factors affecting the decision making, which could include time pressure, the likelihood of increased uncertainty, an increased need for multiagency collaboration, involvement of officials at a higher level, and a strong demand for timely communication.

The main output of these policies in food safety is to define the acceptable level of a microbial hazard, which was previously expressed as a level that is 'As Low As Reasonably Achievable' (ALARA). For many years, the ALARA has been applied in the food industrial sector to guarantee safe food production. The basis of this approach is that if the food industry could improve constantly, the risk will be reduced. However, to meet a specific public health goal a certain level of technical capability and willingness of industry is necessary, and "reasonably achievable" may not be enough to see a real reduction in disease, or it is only linked to disease reduction in a general sense (Todd 2004). Also, ALARA is an ambiguous concept that can differ among countries and industrial sectors, just as occurs with techno-logical capacity (Toyofoku 2006).

The agreement of SPS (WTO 1995) states that members have the right to adopt SPS measures to achieve their self-determined health protection level. This level is defined as 'Appropriate Level of Protection' (ALOP), and it is estimated for the member (country) establishing a sanitary or physosanitary measurement to protect the lives or the health of humans, animals, or plants within its territory. In the context of food safety, an ALOP is a statement of the degree of public health protection that is to be achieved by the food safety systems implemented in a country. Typically, an ALOP would be articulated as a statement related to the disease burden associated with a particular hazard–food combination and its con-sumption in a country. It is often framed in a context of continual improvement in relationship to disease reduction (FAO/WHO 2002).

The ALOP is not the most adequate concept to develop and implant the neces-sary control measurements throughout the food chain (Havelaar et al. 2004). The terms in which the ALOP is expressed do not form part of the 'language' that the industry or the other operators of the food chain use for food safety management (Gorris 2005). Therefore, the ICMSF (2002) proposes the creation of a new concept, the Food Safety Objective (FSO). The ICMSF (2002) defines FSO as 'The maximum frequency and/or concentration of a hazard in a food at the time of consumption that provides or contributes to the appropriate level of protection (ALOP).' The FSO, as an objective, allows large flexibility in designing and implementing control measurements throughout the food chain (Zwietering 2005). The FSO can be understood as a more or less complex system of objectives that industrials and other operators of the food chain use as a criterion to select and develop the most adequate control measures (Fig. 3.1).

The principal strength of the new framework of microbiological food safety management lies in its structure, based on a system of 'quantifiable' objectives, that, first the ICMSF (2002) and later the FAO/WHO (2004), had tried to delimit and define through different concepts. The two first concepts are defined here as proposed by FAO/WHO (2004):

- Performance Objective: 'The maximum frequency and/or concentration of a hazard in a food at a specified step in the food chain before the time of consumption that provides or contributes to an FSO or ALOP, as applicable.'
- Performance Criteria: 'The effect in frequency and/or concentration of a hazard in a food that must be achieved by the application of one or more control measures to provide or contribute to a PO or an FSO.'

These terms and concepts must again be translated to others that food operators can understand, which are the process criteria and product criteria. van Schothorst (2002) defined the process criteria as the control parameters (e.g., time, temperature, pH, a_w), as a step that can be applied to reach an efficiency criteria. In an HACCP context these would correspond with the control limits of a process (Jouve 1999) and are defined as the parameters of a food product that are essential to assure that a FSO will be reached (van Schothorst 2002, 2005). This set of objectives, criteria, and limits can be considered in the HACCP programs and GMP/GHP guides to finally achieve in this way a FSO (van Schothorst 2005).

To determine performance criteria, the inequation proposed by the ICMSF (2002) can be applied. The inequation, in a few words, considers the effect of different processes and sub processes along the food chain (growth, inactivation, cross-contamination, etc.) to reach an FSO:

$$H_0 + \sum I + \sum R \leq \text{FSO} \tag{6.1}$$

H_0 is the initial population of microorganisms, I is a factor of increase, and R is a factor of reduction. The terms are expressed in $\log_{10}$.

In the next sections, the integration of predictive microbiology within these risk management concepts is explained.

6.2   Quantitative Microbial Risk Assessment Needs Predictive Models

Microbial Risk Assessment (MRA) can be defined as a scientific approach intended to estimate the microbial risk associated with food-borne pathogens (Codex Alimentarius Commission 1999). Lammerding and Paoli (1997) pointed out that MRA assesses the impact of changes and trends in the food supply chain; thus, MRA should be able to predict how active or passive changes during processing, distribution, and consumption of foods affect public health. By taking a quantitative approach (i.e., using numerical information), MRA yields more accurate estimations reducing misinterpretation or bias when risk managers use this information. In general, Quantitative MRA (QMRA) is preferred to qualitative MRA, when necessary data and quantitative information are available (Lammerding and Fazil 2000).

QMRA aims to quantitatively describe the effect of food processes, from farm to fork, on microbial risk. To this end, microbial prevalence (e.g., percent of contaminated servings) and concentration (e.g., log cfu/g) along the food chain should be quantified. However, some limitation arises because of the lack of knowledge or information, for example, during pasteurization when pathogen levels are reduced to undetectable levels at which microbiological analysis is not effective or also at the moment of consumption where performing a quantitative analysis is not feasible in a practical sense (Lammerding and Fazil 2000). Thus, predictive models should be incorporated into QMRA to estimate the quantitative effect of determined steps or stages along the food chain for which no data are available or data collection is difficult. Perhaps the work that best summarizes this approach is one of the first published works approaching QMRA and predictive microbiology in a strict sense, which was carried out by Cassin et al. (1998). In this work, different predictive models are applied to estimate the final risk by *Escherichia coli* O157:H7 associated with consumption of beef hamburgers in the U.S. population.

Of the four components constituting MRA methodology, two components require an important contribution of predictive models, that is, hazard characterization and exposure assessment. Hazard characterization is mainly based on applying dose–response models. This type of model establishes a mathematical relationship between ingested dose of the food-borne pathogen and host response in terms of probability of infection, illness, or death (Buchanan et al. 2000). Information from outbreaks, epidemiological studies (Strachan et al. 2005), and experimentation with animal and humans (studies in vivo) are used to derive dose–response models, although important limitations are associated with this kind of studies because of ethical issues, scarcity of information, natural variability, etc. Different types of dose–response model can be developed and applied in QMRA studies; many of these are thoroughly reviewed in the book by Haas et al. (1999). Nonetheless, dose–response modeling still needs important advances to provide more reliable and accurate models. Therefore, new modeling strategies are being proposed, such as the Key Events Dose–Response Framework (KEDRF), which is an analytical approach to model dose–response relationships on the basis of existing information (Buchanan et al. 2009; Julien et al. 2009).

In quantitative exposure assessment, predictive models are intended to describe prevalence and concentration changes at different stages along the food chain (Klapwijk et al. 2000). In other words, the final goal of an exposure assessment study is to know the exposure level to a pathogenic microorganism in terms of prevalence and concentration at the moment of consumption. In this case, predictive microbiology can provide suitable models to better describe different bacterial processes in food-related environments, such as bacterial growth and inactivation (e.g., pasteurization), bacterial transfer (e.g., cross-contamination during handling of foods), and growth probability under determined storage and preservation conditions. Many of the models presented in previous chapters can be now used in a more applicable context to determine the risk associated with certain food(s) and pathogenic microorganism(s). Although important efforts have been made to

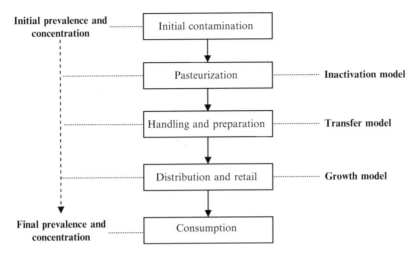

Fig. 6.2 Example of process diagram for an exposure assessment model using specific predictive models for describing microorganism transmission along the represented food chain

agree on a standardized methodology (van Gerwen et al. 2000; FAO/WHO 2008; Bassett et al. 2012), the application of predictive models in QMRA studies is still a complex task, a blend of science and art when information is scarce or data are not conclusive.

Basically, quantitative exposure assessment studies follow a systems approach to represent each food process in the food chain (Lindqvist et al. 2002; Nauta 2003; Pérez-Rodríguez et al. 2006; Carrasco et al. 2010; Tromp et al. 2010). First, a clear representation of the scope of the exposure assessment should be defined according to the requirements or questions to be answered by the QMRA study. Then, the most significant steps in the selected food chain are considered and often represented in a process diagram (Fig. 6.2). For each step, quantitative data or predictive models can be applied to estimate the attendant change of concentration and prevalence of the studied microorganism. Thus, the model output of a previous step would be the model input of a subsequent step in the food chain and so on until the final step is reached in which the final exposure to the pathogen is estimated.

An interesting example of exposure assessment methodology is the Modular Process Risk Model (MPRM) (Nauta 2001, 2003, 2005). This methodology is a development of the process risk model presented by Cassin et al. (1998). Basically, this approach proposes that any step throughout the food chain can be described mathematically through six basic processes: growth, inactivation, mixing, partitioning, removal, and cross-contamination (Table 6.1). In this methodology, a systems approach is taken in which the food chain to be modeled is compartmentalized considering the six basic processes. Through the application of six types of models, an accurate estimation can be made of the level of exposure to a pathogen from a specific food process. According to Nauta (2001), the model of exposure assessment should be divided into small parts, as many as necessary to express the

Table 6.1 Qualitative effect of the six basic processes proposed by Nauta (2001) on prevalence, concentration, and portion size

Process	Effect on the fraction of contaminated units	Effect on the total number cells over all units	Effect on unit size
Growth	=	+	=
Inactivation	−	−	=
Mixing	+	=	+
Partitioning	−	=	−
Removal	−	−	=
Cross-contamination	+	=/+	=

'=' no effect, '+' increase ,'−' decrease

Source: Adapted from European Commission (2003)

logic between the variables and thus model as accurately as possible the relationship between them, all in accordance with the purposes or objectives of the assessment. The main advantage of this method lies in the systematization and simplification of the model creation process.

To estimate the final risk (e.g., probability of becoming ill or number of cases associated with a microorganism and food), the dose–response model and the exposure assessment model should be combined by entering the dose (prevalence and concentration) estimated by the exposure assessment model into the dose–response model equation. Several methodological approaches have been proposed to carry out this mathematical procedure, some of which have been studied in detail in more specific works (Pérez-Rodríguez et al. 2007b, 2011).

In conclusion, QMRA methodology is a relatively recent research area, which requires predictive microbiology models. In spite of the aforementioned advances during the past years, there are still several methodological questions to be answered regarding how models may be applied to obtain a more accurate and precise risk estimation. How rare events (i.e., sporadic cases) can be incorporated in risk assessment studies and how predictive models can help to address this issue are examples of such new challenges faced by risk assessors (Oscar 2011). Therefore, the development of new and innovative predictive models is needed to improve the QMRA studies and fill in the existing and forthcoming data gaps.

6.3 Uncertainty and Variability

6.3.1 Definition of Uncertainty and Variability

Although the terms variability and uncertainty may be easily confounded, they are distinct concepts defined within a decision-making context (NRC 1994). Variability refers to temporal, spatial, or interindividual differences (heterogeneity) in the value of an input (Cullen and Frey 1999). For example, variability might refer to differences in the growth capability between bacterial strains, or in the consumption patterns among consumers. In general, variability cannot be reduced by additional studies or measurements. In contrast, uncertainty is the level of ignorance about an unknown quantity because of analytical limitations or low precision of a measurement system. Therefore, uncertainty can be reduced by increasing the number of analyses or improving measurement precision (Cullen and Frey 1999).

To quantify both components in variables, statistical indexes can be applied, such as standard deviation or coefficient of variation. However, if more detail is required concerning how the variability and uncertainty of a variable behave, probability distributions should be used describing the probability or frequency of occurrence for each possible value of the studied variable. However, the interpretation of the distributions differs in each case. Usually, variability is represented as distributions of frequencies that provide the relative frequency of values from a specific interval. In turn, uncertainty is accounted for by a set of probability distributions, which reflect the degree of belief, or subjective probability that a known value is within a specified interval.

6.3.1.1 Uncertainty and Variability in Predictive Models

As already known, predictive models are not as accurate as would be required by risk managers or, in general, by end-users. Predictions are often far, in a certain extent, from observed microbial response, even though in many instances that is considered admissible. Many factors are responsible for such lack of accuracy, among which we highlight experimental error (e.g., experimenter variability), biological variability (e.g., intra- and interbacterial strain variability), measurement error (e.g., plate counting error), uncontrolled factors (antimicrobial substances in foods), and the food matrix used. Accordingly, prediction models are strongly associated with an important variability and uncertainty component, which should be adequately addressed when models are developed and applied.

During the past few years, some authors have stressed the importance of considering variability and uncertainty when kinetic models are developed (Nauta 2000). In some cases, experimental error derived from plate counting has been incorporated by using probability distributions combined by a Monte Carlo approach (discussed next) (Poschet 2003). With this approach, authors obtained different multiple values for the kinetic parameters (e.g., growth rate) whose variations were described with probability distributions (i.e., normal distribution). In a study by Membré et al. (2005), the variability resulting from the bacterial strains for different bacterial species was studied regarding its effect on growth rate, resulting in estimation of the 95% confidence interval for the obtained growth curves. In this respect, the use of confidence intervals and prediction limits should be further encouraged to better know the uncertainty and variability associated with predictions in addition to facilitating validation process in foods. The use of Bayesian techniques have been also proposed to incorporate variability and uncertainty in growth predictive models (Pouillot et al. 2003). Another important issue to be addressed is the separation between variability and uncertainty caused by its relevance in decision-making processes (Pérez-Rodríguez et al. 2007a), even though, in some situations, to separate variability and uncertainty in the microbial response variation is not sufficiently clear (Nauta 2000). An interesting and illustrative dissertation about this issue was carried out in great detail by Nauta (2007).

6.3.2   Uncertainty and Variability in QMRA

QMRA can be addressed from two different approaches: point-estimate and probabilistic. The first approach concerns use of point-estimate values to describe variables of the model (Lammerding and Paoli 1997). With a deterministic approach only individual scenarios are analyzed, which can correspond to a worst-case, best-case, or average-case scenario depending on the type of point-estimate selected (i.e., average, maximum, minimum value). In the second approach, variables are defined with distributions of probability representing uncertainty and/or variability in variables. Both approaches enable supporting adequate decisions in decision-making processes, although by considering variability and uncertainty, a more complete estimate can be obtained because all possible scenarios are being considered through the use of probability distributions. Hence, an increasing number of probabilistic risk assessment studies has been published during the past few years (Lindqvist et al. 2002; Pouillot et al. 2007; Pérez-Rodríguez et al. 2007b; Tromp et al. 2010). In relationship to these two components, variability and uncertainty, QMRA models can be classified as first- or second-order models (with one or two dimensions, respectively) (Hoffman and Hammonds 1994). First-order or one-dimensional models do not establish a separation between both components, in such a way that variability and uncertainty are expressed by a single probability distribution. Second-order or two-dimensional models, on the other hand, apply more than one distribution for any one variable (Vose 2000). In these cases, variability is explained with the values contained in each probability distribution and uncertainty is considered as a set of probability distributions. Second-order models are preferred as the decision-making process is greatly benefited by this approach (Cox and Ricci 2005). In spite of this, the number of studies on separation between uncertainty and variability is still scarce (Delignette-Müller et al. 2006; Pérez-Rodríguez et al. 2007b; Pouillot and Delignette-Muller 2010; Ottoson et al. 2011), mainly because high-quality data and information are required to discriminate between the two components.

6.3.3   Operating with Probability Distributions

The most used techniques to propagate uncertainty and variability in a probabilistic food risk assessment model comprise classic statistics and numerical methods (Vose 2000). The method of moments is a classical method that can be applied to propagate information about uncertainty and variability on the basis of the properties of mean and standard deviation of input values. However, this method is only valid when input values are distributed normally. By contrast, the algebraic method can be applied even when other types of distributions than the normal distribution are used to characterize uncertainty and variability; however, this method is limited to specific distributions, which are not usually used in risk assessment studies. The Monte Carlo analysis is a numerical method that allows propagating numerous types of probability distributions in risk assessment studies based on the random sampling processes of each distribution. This method has become quite popular among food risk assessors and managers by the existence of commercial software, which enables an easy application by users who are not advanced practitioners in numerical methods.

Chapter 7
Future Trends and Perspectives

Abstract New methodologies have been proposed to be incorporated in predictive microbiology in foods and quantitative microbial risk assessment (QMRA) to achieve more reliable models and facilitate predictive model applications. The meta-analysis is one of the proposed strategies focused on a systematic analysis of a large collection of data with the intention of generating standardized and summarized information to produce a global estimate. This data analysis approach can be applied to better understand the relationship between environmental factors and kinetic parameters or to input QMRA studies to assess the effect of a particular intervention or treatment concerning food safety. The emergence of systems biology is also affecting predictive microbiology, offering new and more mechanistic approaches to yield more reliable and robust predictive models. The so-called genomic-scale models are built on a molecular and genomic basis supported by experimental data obtained from the genomic, proteomic, and metabolomic research areas. Although the existing gene-scale models are promising regarding prediction capacity, they are still few and limited to specific model microorganisms and situations. Further research is needed, in the coming decades, to complete omics information and thus to produce more suitable models to be applied to real-world situations in food safety and quality.

Keywords Meta-analysis • Stepwise process • Data bases • Systems biology • Genomic-scale models • Flux balance analysis (FBA) • Network analysis • Objective function

7.1 Introduction

7.1.1 Meta-Analysis Approach and Benchmarking Data

As previously explained, Quantitative Microbial Risk Assessment (QMRA) is an iterative process that gives insight into setting microbiological criteria and

F. Pérez-Rodríguez and A. Valero, *Predictive Microbiology in Foods*,
SpringerBriefs in Food, Health, and Nutrition 5, DOI 10.1007/978-1-4614-5520-2_7,
© Fernando Pérez-Rodríguez and Antonio Valero 2013

identifying the most relevant factors along the food chain. However, it is recognized that the great amount of data required is the most important drawback to be implemented. Also, as a multidisciplinary area, data processing is becoming more difficult as information is reported in a heterogeneous form. The need to account for variability and uncertainty sources together with the characterization of the main statistical distributions to describe the data leads to the creation of alternative tools to integrate these findings and provide a global estimate. A meta-analysis is a systematic analysis of a large collection of data from individual studies aiming to integrate the information generated in a QMRA study and to produce a global estimate of the effect of a particular intervention or treatment (van Besten and Zwietering 2012). This technique has been more extensively used in food microbiology and can give an improved understanding of main and side effects on microbiological kinetics (Ross et al. 2008).

To start the application of a meta-analysis approach, a sufficient number of data should be generated. Gonzales-Barron and Butler (2011) suggested a stepwise procedure to meta-analysis consisting of (1) systematic review; (2) data extraction to collate quantitative and qualitative information from the primary studies; (3) selection of the appropriate effect size parameter to describe, summarize, and compare the data of the primary studies, and when needed, subsequent translation of the reported findings of the individual studies into the parameter; (4) estimation of the overall effect size by combining the primary studies; (5) assessment of heterogeneity among the studies; and, finally, (6) the presentation of the meta-analysis results.

Selection of data coming from primary studies can begin with experimental data from research institutions or extra data available in scientific data bases. However, individual results must be incorporated into the meta-analysis when they are properly defined, structured, and transparently reported.

In the systematic review process the information to be included in the meta-analysis has to be sufficiently accurate to answer the embedded question of a given case study. For instance, in a lettuce disinfection process, one can measure several heads of lettuce to see if there is contamination by *Escherichia. coli*. The data introduced in the meta-analysis approach should justify if the intervention (disinfection) makes a causal inference on the outcome (presence/absence of *E. coli*) and, if so, how large the effect is.

The data extraction from the primary studies should provide the information necessary for summarizing and synthesizing the results and include both numeric and nonnumeric data.

Effect size refers to the degree to which the phenomenon is present in the population (reduction of *E. coli* numbers by disinfection). For the primary studies, meta-analysis converts the effect size into a 'parameter' that allows direct comparison and summation of the primary studies. There are many types of effect size parameters: (1) binary or dichotomous, for example, indicating the presence or absence of the event of interest in each subject, (2) continuous, and (3) ordinal, where the outcome is measured on an ordered categorical scale.

For the estimation of the overall size effect, primary studies may be weighted to reflect sample size, quality of research design, or other factors influencing their reliability. A relevant factor in precision is the sample size, with larger samples yielding more precise estimates than smaller samples. Another factor affecting precision is the study design, with matched groups yielding more precise estimates (as compared with independent groups) and clustered groups yielding less precise estimates. This consideration can also imply that the obtaining of a lower variance in the primary studies leads to a more accurate global estimate.

On the other hand, it is necessary to make a heterogeneity test among the primary samples to assess the extra-variation in the meta-analysis approach. Generally speaking, individual samples are weighted and statistically compared with aiming at quantifying the variability associated to heterogeneity. In food microbiology, most of the microbial data have been generated in culture media and the effect of environmental factors may not necessarily reflect what might happen in an actual food. Also, results of different studies on factors influencing microbial kinetics are not always similar or may be even contradictory. Variations among microbial strains, individual cell studies, or model estimations contribute positively to increase variability in results. Therefore, quantitative information about the influence of various factors on microbial kinetics is often not adequate under specific conditions, and also often is not available in the published literature.

Finally, results coming from the meta-analysis are presented into several graph types, such as bubble plots, which display point estimates and confidence intervals of each primary study and the overall effects in the global estimate.

The use of data bases in predictive microbiology can provide thousands of records of microbial growth or inactivation kinetics under a wide range of environmental conditions. A systematic and critical analysis of the literature followed by integration of the gathered data results in global estimates of kinetic parameters with their variability, and these can be used to benchmark the latest published data (van Asselt and Zwietering 2006). Meta-analysis has been used in various QMRA studies for relating the microbial concentration of a given hazard to a public health outcome (Pérez-Rodríguez et al. 2007b). However, large variability sources are expected in some cases, mainly because of heterogeneity in primary data. Additionally, overlapping problems are generated when the same information of one variable is obtained from different studies. In spite of these disadvantages, when a large dataset is manipulated, meta-analysis can provide useful links to discern between explanatory variables on the global estimate. The construction of updated data bases on the reviewed question or parameter can also reveal the present knowledge, can highlight default areas where there is a lack of information on factors that might affect the parameter of interest, and can therefore provide direction for future research.

7.2 Mechanistic Predictive Models

Advances in molecular biology, particularly in genome sequencing and high-throughput measurements, enable us to obtain comprehensive data on the cellular system and gain information on the underlying molecules (Kitano 2002).

This genomics revolution has in the past years provided researchers with the option to look genome wide for cellular responses at the level of gene expression (Keijser et al. 2007) and protein presence (Wolff et al. 2006; Hahne et al. 2010). The need of integrating all this complex information has contributed to an emerging scientific field, so-called systems biology, aimed at understanding complex biological systems at the systems level (Kitano 2001). The fundamental idea behind the systems biology approach is that biological systems are hierarchically organized with influences going both up and down through the hierarchy (Brul et al. 2008).

The great avalanche of 'omics' data (i.e., genomic and proteomic data) in systems biology necessitates applying mathematical methods to better understand the interactions and relationships among the different elements within the studied system (Fig. 7.1). Stelling (2004) classified mathematical models applied in systems biology in interaction-based models, constraint-based models, and mechanism-based models. The interaction-based models refer to network topology analysis in which interactions between the different elements in the system, for example, metabolic reactions, protein–protein interactions, and gene regulation, are accounted for by graphical networks. In constraint-based models, physicochemical properties such as reaction stoichiometries and reversibilities impose constraints on network function in addition to network topology. This network reconstruction process ultimately results in the generation of a biochemically, genomically, and genetically (BiGG) structured data base that can be further utilized for both mathematical computation and analysis of high-throughput data sets. The network spans the set of metabolic reactions taking place in a specific biological system, assuming a stationary state (Hertog et al. 2011) in which each reaction is referred to as a flux. The methodologies developed in metabolic engineering such as metabolic control analysis and metabolic flux analysis are applied to analyze steady-state fluxes, although these may also be used to explain oscillatory systems so long as average fluxes are considered (Schuster et al. 2002). More quantitative models can be addressed based on kinetic rates of metabolic reactions included in the biological networks. In this approach, a system of linear differential equations is used to account for reactions rate of the quasi-dynamic or dynamic state fluxes (Hertog et al. 2011). As new genomic data become available, these may aid in the parameterization of metabolic models (Voit 2002). However, one weakness of this approach is that it ignores the variability and noise found in biological networks, which may have important implications in their function (Heath and Kavraki 2009). To overcome this limitation, a stochastic approach has been proposed that basically consists of adding a noise term to the differential equations. Similarly, gene expression regulation (i.e., transcription and translation) and signaling networks have a probabilistic nature that should be accounted for by applying a stochastic approach (Treviño Santa Cruz et al. 2005; McAdams and Arkin 1997).

The latter type of model mentioned by Stelling (2004) refers to mechanism-based models. The author means that with this type of model, models can predict the system dynamics by integrating detailed mechanisms operating in metabolism, signal processing, and gene regulation. The success of this mechanistic approach,

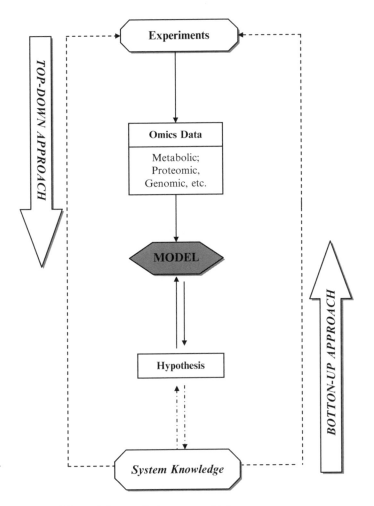

Fig. 7.1 Scheme of the workflow applied to systems biology

that is, integrated modeling, relies largely on the availability of information about the complete mechanism and attendant parameters.

Traditionally, in the field of predictive microbiology applied to foods, the scarce information on the mechanisms involved in the cellular functions has hampered microbiologists from undertaking more mechanistic models, albeit some mechanistic parameters has been introduced in specific cases (Baranyi and Roberts 1995). The emergence of systems biology is creating a new path for microbiologists in predictive microbiology, offering new and more mechanistic approaches to give rise to more reliable and robust models (Brul et al. 2008). In so doing, predictive microbiology will be able to move from the most used empirical modeling, that is, black box models, toward so-called white or gray box models, based on an better understanding of the biological functions in cells, enabling providing more accurate predictions under specific physical and chemical changes and even extending the model outside the range of space bounded by observations. Such data not only allow for a better fine-tuning of growth/no growth boundaries but will also begin to strengthen die-off/survival models (Brul et al. 2008). Several computer models have been developed on the basis of information derived from systems biology studies and wealthy databases. However, many of the mechanistic studies have been done under conditions and in model microorganisms with relatively low practical relevance (Brul et al. 2008). One of the most studied microorganisms is *E. coli*, as much is known about its metabolism, regulation, and genome, enabling the development of more mechanistic and reliable in silico models for this model microorganism (Reed and Palsson 2003). The experience obtained with *E. coli* has served to be applied to other microorganisms such as *Haemophilus influenzae* (Edward and Palsson 1999), *Helicobacter pylori*, and *Saccharomyces cerevisiae* (Petranovic and Vemuri 2009).

To date, few systems biology-based models have been explored or developed within the area of predictive microbiology in foods (Brul et al. 2008). However, constraint-based models seem to be the first choice by microbiologists to understand the behavior of microorganisms in food-related environments (Métris et al. 2011; Peck et al. 2011). The most significant kinetic reactions constituting the metabolism of the model bacterium are modeled and simulated to know which specific metabolic processes are related to a determined bacterial response (e.g., outgrowth, adaptation, survival). These models consist of describing the fluxes that make up a metabolic network in which each flux accounts for a metabolic reaction as concentration change per time unit for the substrate and product. The reactions can be described by a system of linear differential equations in which stoichiometric coefficient of equations are assumed to be constant because the model represents

$$\frac{dx}{dt} = Sv = 0 \tag{7.1}$$

Here, x defines a vector of the intermediate concentrations of metabolites at a specific time, S is the stoichiometric matrix describing all the metabolic reactions, and S_{ij} corresponds to the ith stoichiometric coefficient in the jth reaction. The thermodynamic constraints and enzyme capacity constraints are represented by vector $v = [v_1, \ldots v_j]$, which includes the reaction rates of each metabolic reaction or flux. Setting Eq. (7.1) to 0 means that conservation laws apply in the production and consumption rates (i.e., $rate_{consumption} = rate_{production}$). A simplified example of the steady-state flux might be the well-known coenzyme nicotinamide adenine dinucleotide (NAD), involved in many metabolic routes as an electron donor. In this case, the reaction would be

$$NAD^+ + H \rightarrow NADH$$

According to the law of conservation, NAD + NADH = a constant, which means that the sum of concentrations of NAD and NADH does not change with time.

The derivation of the reaction rate equations is another important aspect and should be based on an appropriate metabolic network, which should be completely known and closed. The quasi-steady-state and rapid equilibrium approaches can be used to obtain the reaction rate equations. With regard to the kinetic parameters, these might be estimated by using sources such as literature data, electronic data bases, experimental data for dependencies between initial reactions rates and products, inhibitors, substrates, and activators, and finally time-series data for enzyme kinetics and whole pathways (Demin et al. 2005).

Because the system of equations has more fluxes than metabolites, the system is underdetermined (Kauffman et al. 2003), which means that the system has multiple solutions. To reduce the solution space of the system, the model is constrained by imposing different rules, which are often related to thermodynamic feasibility, enzymatic capacity, and mass balance. Model solutions that do not comply with such criteria are excluded from the solution space of the model (Reed and Palsson 2003). Once constraints are defined for the model, the corresponding solution space should be determined. To this end, several mathematical approaches can be taken such as linear optimization, elementary modes and extreme pathways, phenotypic phase plane analysis, gene deletions, or finding objective functions. The linear optimization, which is referred to as flux balance analysis (FBA), is based on an objective function, which is utilized to define the solution space by maximizing or minimizing the defined objective function (Feist and Palsson 2010; Varma and Palsson 1994). The most used objective functions include ATP production, production of a specific by-product, and biomass production (i.e., growth rate) (Van Impe et al. 2011; Reed and Palsson 2003). In this respect, using a biomass production objective function can accurately estimate the growth rate of E. coli, as evidenced by the work by Feist et al. (2007).

Métris et al. (2011) performed in silico simulations based on the model of E. coli K12 MG1655 previously developed by Feist et al. (2007) considering 1,387 metabolic reactions and 1,260 genes. This study can be considered as one of the first approaches of predictive microbiology in the foods area to systems biology modeling. This model applied the most often used objective function based on optimizing the biomass production, which is associated with growth-associated maintenance (GAM) energy and non-growth-associated maintenance (NGAM) energy. Their values were derived from experiments in a chemostat without added NaCl, which we refer to as the control conditions (Feist et al. 2007). The model was modified to consider exposure to osmotic stress by including changes of concentrations in an osmoprotectant associated with osmolarity changes. The work did not find definitive results relating the changes of these substances with a decrease of the growth rate. Similarly, the model was tested to ascertain if biomass composition derived from osmotic stress might explain the decrease of growth rates observed in experiments; however, again the results were not conclusive. Finally, the authors suggested that more specific objective functions should be developed to explain the chemicophysical limitations of the growth rate. For that, authors suggested including gene regulation, crowding, and other additional cell resources such as ribosomal content and some tradeoff observed under osmotic stress. This work and its results provide evidence that a new modeling approach is emerging, although still with important gaps and limitations. Nonetheless, it might provide the necessary theoretical basis to develop more mechanistic predictive models in foods (Fig. 7.1).

Chapter 3
Predictive Models: Foundation, Types, and Development

F. Pérez-Rodríguez and A. Valero, *Predictive Microbiology in Foods*,
SpringerBriefs in Food, Health, and Nutrition 5, DOI 10.1007/978-1-4614-5520-2, pp. 25–55,
© Fernando Pérez-Rodríguez and Antonio Valero 2013

The publisher regrets the following errors:

The presentation of Equation 3.8 is incorrect:

$$h_0 = \ln\left(1 + \frac{1}{q_0}\right) = \mu_{\max}\lambda \qquad (3.8)$$

The correct equation is:

$$h_0 = \ln\left(1 + \frac{1}{q_0}\right) = \mu_{\max}t_{lag} \qquad (3.8)$$

The presentation of Equation 3.9 is incorrect:

$$\begin{cases} \ln(N) = \ln(N_0), & t \leq lag \\ \ln(N) = \ln(N\,max) - \ln\left[1 + \left(\dfrac{N\,max}{N_0} - 1\right)\exp(-\mu\,max(tlag - lag))\right], & t > lag. \end{cases}$$

(3.9)

The corrected equation is:

$$\begin{cases} \ln(N) = \ln(N_0), & t \leq t_{lag} \\ \ln(N) = \ln(N_{max}) - \ln\left[1 + \left(\dfrac{N_{max}}{N_0} - 1\right)\exp(-\mu_{max}(t - t_{lag}))\right], & t > t_{lag} \end{cases}$$

(3.9)

The presentation of Equation 3.40 is incorrect:

$$LnK_{max} = C_0 + \frac{C_1}{T} + C_2 pH + C_3 pH^2 + C_4 a_w^2$$

(3.40)

The corrected equation is:

$$Ln\,k_{max} = C_0 + \frac{C_1}{T} + C_2 pH + C_3 pH^2 + C_4 aw^2$$

(3.40)

DOI 10.1007/978-1-4614-5520-2_8

The online version of the original chapter can be found at
http://dx.doi.org/10.1007/978-1-4614-5520-2_3

References

Abdi H (2007) The method of least squares. In: Salkind N (ed) Encyclopedia of measurements and statistics. Sage, London

Abramoff MD, Magalhaes PJ, Ram SJ (2004) Image processing with Image J. J Biophotonics Int 11:36–42

Akaike H (1974) A new look at the statistical model identification. IEEE T Automat Contr 19:716–723

Almeida JS (2002) Predictive non-linear modeling of complex data by artificial neural networks. Curr Opin Biotechnol 13:72–76

Antwi M, Bernaerts K, Van Impe JF, Geeraerd AH (2007) Modelling the combined effects of structured food model system and lactic acid on *Listeria innocua* and *Lactococcus lactis* growth in mono- and co-culture. Int J Food Microbiol 120:71–84. doi:10.1016/j.ijfoodmicro. 2007.04.015

Arroyo-López F, Orlić S, Querol A, Barrio E (2009) Effects of temperature, pH and sugar concentration on the growth parameters of *Saccharomyces cerevisiae*, *S. kudriavzevii* and their interspecific hybrid. Int J Food Microbiol 131:120–127. doi:10.1016/j.ijfoodmicro. 2009.01.035

Augustin JC, Carlier V (2000a) Modelling the growth rate of *Listeria monocytogenes* with a multiplicative type model including interactions between environmental factors. Int J Food Microbiol 56:53–70. doi:10.1016/S0168-1605(00)00224-5

Augustin JC, Carlier V (2000b) Mathematical modelling of the growth rate and lag time for *Listeria monocytogenes*. Int J Food Microbiol 56:29–51. doi:10.1016/S0168-1605(00)00223-3

Augustin JC, Rosso L, Carlier V (1999) Estimation of temperature dependent growth rate and lag time of *Listeria monocytogenes* by optical density measurements. J Microbiol Methods 38:137–146

Augustin JC, Rosso L, Carlier V (2000) A model describing the effect of temperature history on lag time for *Listeria monocytogenes*. Int J Food Microbiol 57:169–181. doi:10.1016/S0168-1605(00)00260-9

Aziza F, Mettler E, Daudin JJ, Sanaa M (2006) Stochastic, compartmental, and dynamic modeling of cross-contamination during mechanical smearing of cheeses. Risk Anal 26:731–745. doi:10.1111/j.1539-6924.2006.00758.x

Baert K, Valero A, De Meulenaer B, Samapundo S, Ahmed MM, Bo L, Debevere J, Devlieghere F (2007) Modeling the effect of temperature on the growth rate and lag phase of *Penicillium expansum* in apples. Int J Food Microbiol 118:139–150. doi:10.1016/j.ijfoodmicro.2007. 07.006

Baranyi J (1992) Letters to the editor: A note on reparameterization of bacterial growth curves. Food Microbiol 9:169–171. doi:10.1016/0740-0020(92)80024-X

F. Pérez-Rodríguez and A. Valero, *Predictive Microbiology in Foods*,
SpringerBriefs in Food, Health, and Nutrition 5, DOI 10.1007/978-1-4614-5520-2,
© Fernando Pérez-Rodríguez and Antonio Valero 2013

Baranyi J (1998) Comparison of stochastic and deterministic concepts of bacterial lag. J Theor Biol 192:403–408. doi:10.1006/jtbi.1998.0673

Baranyi J, Pin C (2001) A parallel study on bacteria growth and inactivation. J Theor Biol 210:327–336. doi:10.1006/jtbi.2001.2312

Baranyi J, Roberts TA (1994) A dynamic approach to predicting bacterial growth in food. Int J Food Microbiol 23:277–294. doi:10.1016/0168-1605(94)90157-0

Baranyi J, Roberts TA (1995) Mathematics of predictive food microbiology. Int J Food Microbiol 26:199–218

Baranyi J, Tamplin ML (2004) ComBase: a common database on microbial responses to food environments. J Food Prot 67:1967–1971

Baranyi J, Roberts TA, McClure P (1993) A non-autonomous differential equation to model bacterial growth. Food Microbiol 10:43–59. doi:10.1006/fmic.1993.1005

Baranyi J, Robinson TP, Kaloti A, Mackey BM (1995) Predicting growth of *Brochothrix thermosphacta* at changing temperature. Int J Food Microbiol 27:61–75. doi:10.1016/0168-1605(94)00154-X

Baranyi J, Ross T, McMeekin T, Roberts TA (1996) The effect of parameterisation on the performance of empirical models used in Predictive Microbiology. Food Microbiol. 13:83–91. DOI: 10.1006/fmic.1996.0011

Baranyi J, Pin C, Ross T (1999) Validating and comparing predictive models. Int J Food Microbiol 48:159–166. doi:10.1016/S0168-1605(99)00035-5

Basheer I, Hajmeer M (2000) Artificial neural networks: fundamentals, computing, design and application. J Microbiol Methods 43:3–31. doi:10.1016/S0167-7012(00)00201-3

Bassett J, Nauta M, Zwietering MH (2012) Tools for microbiological risk assessment. ILSI Europe Series 2012:1–40

Begot C, Desnier I, Daudin JD, Labadie JC, Lebert A (1996) Recommendations for calculating growth parameters by optical density measurements. J Microbiol Methods 25:225–232. doi:10.1016/0167-7012(95)00090-9

Bernaerts K, Dens E, Vereecken K et al (2004) Concepts and tools for predictive modeling of microbial dynamics. J Food Prot 67:2041–2052

Bigelow WD (1921) The logarithmic nature of thermal death time curves. J Infect Dis 27:528–536. doi:10.1093/infdis/29.5.528

Bigelow WD, Esty JR (1920) The thermal death point in relation to typical thermophylic organisms. J Infect Dis 27:602–617. doi:10.1093/infdis/27.6.602

Billo EJ (2007) Excel for scientists and engineers: numerical methods. Wiley, Hoboken

Blackburn CW, Curtis LM, Humpheson L, Billon C, McClure PJ (1997) Development of thermal inactivation models for *Salmonella enteritidis* and *Escherichia coli* O157:H7 with temperature, pH and NaCl as controlling factors. Int J Food Microbiol 38:31–44. doi:10.1016/S0168-1605(97)00085-8

Bredholt S, Nesbakken T, Holck A (1999) Protective cultures inhibit growth of *Listeria monocytogenes* and *Escherichia coli* O157:H7 in cooked, sliced, vacuum- and gas-packaged meat. Int J Food Microbiol 53:43–52. doi:10.1016/S0168-1605(99)00147-6

Brown AM (2001) A step-by-step guide to non-linear regression analysis of experimental data using a Microsoft Excel spreadsheet. Comput Methods Prog Biol 65:191–200. doi:10.1016/S0169-2607(00)00124-3

Brul S, Mensonides FIC, Hellingwerf KJ, Teixeira de Mattos MJ (2008) Microbial systems biology: new frontiers open to predictive microbiology. Int J Food Microbiol 128:16–21. doi:10.1016/j.ijfoodmicro.2008.04.029

Buchanan RL (1992) Predictive microbiology. Mathematical of microbial growth in foods. Food Saf Assess 484:250–260

Buchanan R (1999) Microbial competition: effect of *Pseudomonas fluorescens* on the growth of *Listeria monocytogenes*. Food Microbiol 16:523–529. doi:10.1006/fmic.1998.0264

Buchanan RL, Golden MH (1995) Model for the non-thermal inactivation of *Listeria monocytogenes* in a reduced oxygen environment. Food Microbiol 12:203–212. doi:10.1016/S0740-0020(95)80099-9

Buchanan RL, Klawitter LA (1991) Effect of temperature history on the growth of *Listeria monocytogenes* Scott A at refrigeration temperatures. Int J Food Microbiol 12:235–246. http://dx.doi.org/10.1016/0168-1605(91)90074-Y

Buchanan RL, Whiting RC, Damert WC (1997) When is simple good enough: a comparison of the Gompertz, Baranyi, and three-phase linear models for fitting bacterial growth curves. Food Microbiol 14:313–326. doi:10.1006/fmic.1997.0125

Buchanan RL, Smith JL, Long W (2000) Microbial risk assessment: dose–response relations and risk characterization. Int J Food Microbiol 58:159–172. doi:10.1016/S0168-1605(00)00270-1

Buchanan RL, Havelaar AH, Smith MA, Whiting RC, Julien E (2009) The Key Events Dose–Response Framework: its potential for application to foodborne pathogenic microorganisms. CRC Rev Food Sci 49:718–728. doi:10.1080/10408390903116764

Campos DT, Marks BP, Powell MR, Tamplin ML (2005) Quantifying the robustness of a broth-based *Escherichia coli* O157: H7 growth model in ground beef. J Food Prot 68:2301–2309

Carpentier B, Chassaing D (2004) Interactions in biofilms between *Listeria monocytogenes* and resident microorganisms from food industry premises. Int J Food Microbiol 97:111–122. doi:10.1016/j.ijfoodmicro.2004.03.031

Carrasco E, Pérez-Rodríguez F, Valero A, García-Gimeno RM, Zurera G (2010) Risk assessment and management of *Listeria monocytogenes* in ready-to-eat lettuce salads. Compr Rev Food Sci Food Saf 9:498–512. doi:10.1111/j.1541-4337.2010.00123.x

Cassin MH, Lammerding AM, Todd ECD, Ross W, McColl RS (1998) Quantitative risk assessment for *Escherichia coli* O157:H7 in ground beef hamburgers. Int J Food Microbiol 41:21–44. doi:10.1016/S0168-1605(98)00028-2

Castillejo-Rodríguez AM, Gimeno RMG, Cosano GZ, Alcalá EB, Pérez MR (2002) Assessment of mathematical models for predicting *Staphylococcus aureus* growth in cooked meat products. J Food Prot 65:659–665

Cerf O, Davey KR, Sadoudi AK (1996) Thermal inactivation of bacteria. A new predictive model for the combined effect of three environmental factors: temperature, pH and water activity. Food Res Int 29:219–226. doi:10.1016/0963-9969(96)00039-7

Chatterjee S, Hadi AS (2006) The problem of correlated errors. Regression analysis by example. Wiley Series in Probability and Statistics, 4th edn. Wiley, New York, pp 197–219. ISBN-I3 978-0-471-74696-6

Chen Y, Jackson KM, Chea FP, Schaffner DW (2001) Quantification and variability analysis of bacterial cross-contamination rates in common food service tasks. J Food Prot 64:72–80

Cheroutre-Vialette M, Lebert A (2002) Application of recurrent neural network to predict bacterial growth in dynamic conditions. Int J Food Microbiol 73:107–118. doi:10.1016/S0168-1605(01)00642-0

Christensen BB, Rosenquist H, Sommer HM, Nielsen NL, Fagt S, Andersen NL, Nørrung B (2005) A model of hygiene practices and consumption patterns in the consumer phase. Risk Anal 25:49–60. doi:10.1111/j.0272-4332.2005.00566.x

Codex Alimentarius Commission (1999) Principles and guidelines for the conduct of microbiological risk assessment. CAC/GL-30-1999. Secretariat of the Joint FAO/WHO Food Standards Programme. FAO, Rome

Codex Alimentarius Commission (2007) Principles and guidelines for the conduct of microbiological risk management (MRM). Document CAC/GL 63-2007. FAO, Rome. www.codexalimentarius.net/download/standards/10741/CXG_063e.pdf. Accessed 20 Mar 2012

Cornu M, Kalmokoff M, Flandrois JP (2002) Modelling the competitive growth of *Listeria monocytogenes* and *Listeria innocua* in enrichment broths. Int J Food Microbiol 73:261–274

Cornu M, Billoir E, Bergis H et al (2011) Modeling microbial competition in food: application to the behavior of *Listeria monocytogenes* and lactic acid flora in pork meat products. Food Microbiol 28:639–647. doi:10.1016/j.fm.2010.08.007

Corradini MG, Peleg M (2007) A Weibullian model for microbial injury and mortality. Int J Food Microbiol 119:319–328. doi:10.1016/j.ijfoodmicro.2007.08.035

Cox L, Ricci PF (2005) Causation in risk assessment and management: models, inference, biases, and a microbial risk-benefit case study. Environ Int 31:377–397. doi:10.1016/j.envint.2004. 08.010

Cullen AC, Frey HC (1999) Probabilistic techniques in exposure assessment. A handbook for dealing with variability and uncertainty in models and inputs. Plenum, New York

Cuppers HGM, Smelt JPPM (1993) Time to turbidity measurement as a tool for modeling spoilage by *Lactobacillus*. J Ind Microbiol 12:168–171. doi:10.1007/BF01584186

Dalgaard P, Koutsoumanis K (2001) Comparison of maximum specific growth rates and lag times estimated from absorbance and viable count data by different mathematical models. J Microbiol Methods 43:183–196. doi:10.1016/S0167-7012(00)00219-0

Dalgaard P, Ross T, Kamperman L, Neumeyer K, McMeekin TA (1994) Estimation of bacterial growth rates from turbidimetric and viable count data. Int J Food Microbiol 23:391–404. doi:10.1016/0168-1605(94)90165-1

Dalgaard P, Buch P, Silberg S (2002) Seafood Spoilage Predictor: development and distribution of a product specific application software. Int J Food Microbiol 73:343–349. doi:10.1016/S0168-1605(01)00670-5

Dantigny P, Marín S, Beyer M, Magan N (2007) Mould germination: data treatment and modelling. Int J Food Microbiol 114:17–24. doi:10.1016/j.ijfoodmicro.2006.11.002

Davey KR (1993) Linear-Arrhenius models for bacterial growth and death and vitamin denaturations. J Ind Microbiol 12:172–179. doi:10.1007/BF01584187

Delignette-Müller ML, Cornu M, Pouillot R, Denis JB (2006) Use of Bayesian modelling in risk assessment: application to growth of *Listeria monocytogenes* and food flora in cold-smoked salmon. Int J Food Microbiol 106:195–208. doi:10.1016/j.ijfoodmicro.2005.06.021

Demin OV, Plyusnina TY, Lebedeva GV, Zobova EA, Metelkin EA, Kolupaev AG, Goryanin II, Tobin F (2005) Kinetic modelling of the *Escherichia coli* metabolism. In: Alberghina L, Westerhoff HV (eds) Topics in current genetics, vol 13. Systems biology. Springer, Berlin, pp 31–67

Den Aantrekker ED, Boom RM, Zwietering MH, van Schothorst M (2003a) Quantifying recontamination through factory environments–a review. Int J Food Microbiol 80:117–130. doi:10.1016/S0168-1605(02)00137-X

Den Aantrekker ED, Beumer RR, van Gerwen SJC, Zwietering MH, van Schothorst M, Boom RM (2003b) Estimating the probability of recontamination via the air using Monte Carlo simulations. Int J Food Microbiol 87:1–15. doi:10.1016/S0168-1605(03)00041-2

Dens EJ, Van Impe JF (2000) On the importance of taking space into account when modeling microbial competition in structured food products. Math Comput Simul 53:443–448. doi:10.1016/S0378-4754(00)00239-1

Devlieghere F (2000) Predictive modelling of the spoilage and the microbial safety of modified atmosphere packaged cooked meat products. Ph.D. thesis, Faculty of Agricultural and Applied Biological Sciences, Ghent University, Belgium

Devlieghere F, Geeraerd H, Versyck KJ, Vandewaetere B, Van Impe J, Debevere J (2001) Growth of *Listeria monocytogenes* in modified atmosphere packed cooked meat products: a predictive model. Food Microbiol 18:53–66. doi:10.1006/fmic.2000.0378

US Food and Drug Administration (2009) Microbiological challenge testing. Evaluation and definition of potentially hazardous foods. Available at http://www.fda.gov/Food/Science Research/ResearchAreas/SafePracticesforFoodProcesses/ucm094154.htm. Accessed 20 March 2012

FDA (US Food and Drug Administration) (2009) Chapter 6. Microbiological challenge testing evaluation and definition of potentially hazardous foods. http://www.fda.gov/Food/ScienceResearch/ResearchAreas/SafePracticesforFoodProcesses/ucm094154.htm. Accessed 12 Jan 2012

Dupont C, Augustin JC (2009) Influence of stress on single-cell lag time and growth probability for *Listeria monocytogenes* in half Fraser broth. Appl Environ Microbiol 75:3069–3076. doi:10.1128/AEM.02864-08

Dykes G (1999) Image analysis of colony size for investigatin sublethal injury in *Listeria monocytogenes*. J Rap Methods Auto Microbiol 7:223–231

Dym CL (2004) Principles of mathematical modeling. Elsevier Academic Press, London, 4

Edward JS, Palsson BO (1999) Systems properties of the *Haemophilus influenzae* Rd metabolic genotype. J Biol Chem 274:17410–17416

Endo H, Nagano Y, Ren H, Hayashi T (2001) Rapid enumeration of bacteria grown on surimi-based products by flow cytometry. Fish Sci 67:969–974. doi:10.1046/j.1444-2906.2001.00348.x

Esty JR, Meyer KF (1922) The heat resistance of spores of *B. botulinus* and related anaerobes. J Infect Dis 31:650–663. doi:10.1093/infdis/31.6.650

European Commission (2003) Risk assessment of food borne bacterial pathogens: quantitative methodology relevant for human exposure assessment (final report). ec.europa.eu/food/fs/sc/ssc/out308_en.pdf. Accessed 2 Jan 2012

FAO/WHO (Food Agriculture Organization/World Health Organization) (1995) Application of risk analysis to food standards. Report of the joint FAO/WHO expert consultation. FAO/WHO, Geneva

FAO/WHO (Food Agriculture Organization/World Health Organization) (1997) Risk management and food safety. Report of a joint FAO/WHO consultation. Rome, Italy, 27–31 June 1997. FAO/WHO, Rome and Geneva

FAO/WHO (Food Agriculture Organization/World Health Organization) (1998) Application of risk communication to food standards and food safety. Report of the joint FAO/WHO expert consultation. FAO/WHO, Rome

FAO/WHO (Food Agriculture Organization/World Health Organization) (2001) Codex Alimentarius Commission. Procedural manual, twelfth Edition. Procedures for the elaboration of codex standards and related texts. Joint FAO/WHO food standards programme, codex alimentarius commission. FAO/WHO, Rome. http://www.fao.org/docrep/005/Y2200E/y2200e04.htm#bm04. Accessed 10 Mar 2012

FAO/WHO (Food Agriculture Organization/World Health Organization) (2002) Proposed draft principles and guidelines for incorporating microbiological risk assessment in the development of food safety standards, guidelines and related text. Report of a joint FAO/WHO consultation, Kiel, 18–22 Mar 2002. FAO/WHO and Institute for Hygiene and Food Safety of the Federal Dairy Research Centre, Rome, Geneva and Kiel

FAO/WHO (Food Agriculture Organization/ World Health Organization) (2004). Definitions on food safety objective, performance objective and performance criterion. Joint FAO/WHO food standards programme. Codex committee on food of the hygiene. Thirty-sixth session. Alinorm 04/27/13. FAO/WHO, Rome and Geneva.

FAO/WHO (Food Agriculture Organization/World Health Organization) (2006) Food safety risk analysis: a guide for National Food Safety Authorities. FAO Food and Nutrition Papers-87, FAO, Rome. http://www.who.int/foodsafety/publications/micro/riskanalysis06/en/. Accessed 24 Mar 2012

FAO/WHO (Food Agriculture Organization/World Health Organization) (2008) Exposure assessment of microbiological hazards in food. Microbiological Risk Assessment Series 7

FAO/WHO (Food Agriculture Organization/World Health Organization) (2011) Guide for application of risk analysis principles and procedures during food safety emergencies. Food and Agriculture Organization of the United Nations and World Health Organization, Rome. ISBN 978 92 4 150247 4

Fehlhaber K, Krüger G (1998) The study of *Salmonella enteritidis* growth kinetics using rapid automated bacterial impedance technique. J Appl Microbiol 84:945–949. doi:10.1046/j.1365-2672.1998.00410.x

Feist AM, Palsson BØ (2010) The biomass objective function. Curr Opin Microbiol 13:344–349. doi:10.1016/j.mib.2010.03.003

Feist AM, Henry CS, Reed JL, Krummenacker M, Joyce AR, Karp PD, Broadbelt LJ, Hatzimanikatis V, Palsson BO (2007) A genome-scale metabolic reconstruction for *Escherichia coli* K-12 MG1655 that accounts for 1260 ORFs and thermodynamic information. Mol Syst Biol 3:121. doi:10.1038/msb4100155

Fernandes RL, Nierychlo M, Lundin L, Pedersen AE, Tellez PEP, Dutta A, Carlquist M, Bolic A, Schapper D, Brunetti AC, Helmark S, Heins AL, Jensen AD, Nopens I, Rottwitt K, Szita N, van Elsas JD, Nielsen PH, Martinussen J, Sorensen SJ, Lantz AE, Gernaey KV (2011) Experimental methods and modeling techniques for description of cell population heterogeneity. Biotechnol Adv 29:575–599. doi:10.1016/j.biotechadv.2011.03.007

Fernández PS, Ocio MJ, Rodrigo F, Rodrigo M, Martínez A (1996) Mathematical model for the combined effect of temperature and pH on the thermal resistance of *Bacillus stearothermophilus* and *Clostridium sporogenes* spores. Int J Food Microbiol 32:225–233. doi:10.1016/0168-1605(96)01118-X

Fernandez PS, George SM, Sills CC, Peck MW (1997) Predictive model of the effect of CO_2, pH, temperature and NaCl on the growth of *Listeria monocytogenes*. Int J Food Microbiol 37:37–45. doi:10.1016/S0168-1605(97)00043-3

Ferrer J, Prats C, López D, Vives-Rego J (2009) Mathematical modelling methodologies in predictive food microbiology: a SWOT analysis. Int J Food Microbiol 134:2–8. doi:10.1016/j.ijfoodmicro.2009.01.016

Francois K, Devlieghere F, Standaert AR, Geeraerd AH, Van Impe JF, Debevere J (2003) Modelling the individual cell lag phase. Isolating single cells: protocol development. Lett Appl Microbiol 37:26–30. doi:10.1046/j.1472-765X.2003.01340.x

Francois K, Devlieghere F, Standaert AR, Geeraerd AH, Van Impe JF, Debevere J (2006) Effect of environmental parameters (temperature, pH and a_w) on the individual cell lag phase and generation time of *Listeria monocytogenes*. Int J Food Microbiol 108:326–335. doi:10.1016/j.ijfoodmicro.2005.11.017

Garcia D, Ramos AJ, Sanchis V, Marín S (2010) Modelling mould growth under suboptimal environmental conditions and inoculum size. Food Microbiol 27:909–917. doi:10.1016/j.fm.2010.05.015

García-Gimeno RM, Hervás C, de Silóniz MI (2002) Improving artificial neural networks with a pruning methodology and genetic algorithms for their application in microbial growth prediction in food. Int J Food Microbiol 72:19–30. doi:10.1016/S0168-1605(01)00608-0

Garthright WE (1997) The three-phase linear model of bacterial growth: a response. Food Microbiol 14:395–397. doi:10.1006/fmic.1996.9997

Geeraerd AH, Herremans CHML, Herremans ML, Cenes C, Van Impe JF (1998) Application of artificial neural networks as a non linear technique to describe bacterial growth in chilled food products. Int J Food Microbiol 44:49–68. doi:10.1016/S0168-1605(98)00127-5

Geeraerd AH, Herremans CH, Van Impe JF (2000) Structural model requirements to describe microbial inactivation during a mild heat treatment. Int J Food Microbiol 59:185–209. doi:10.1016/S0168-1605(00)00362-7

Geeraerd AH, Valdramidis VP, Devlieghere F, Bernaert H, Debevere J, Van Impe JF (2004) Development of a novel approach for secondary modelling in predictive microbiology: incorporation of microbiological knowledge in black box polynomial modelling. Int J Food Microbiol 91:229–244. doi:10.1016/S0168-1605(03)00388-X

Geeraerd AH, Valdramidis VP, Van Impe JF (2005) GInaFiT, a freeware tool to assess non-log-linear microbial survivor curves. Int J Food Microbiol 102:95–105. doi:10.1016/j.ijfoodmicro.2004.11.038

Genigeorgis CA (1981) Factors affecting the probability of growth of pathogenic microorganisms in foods. J Am Vet Med Assoc 179:1410–1417

Gibson AM, Bartchetll N, Roberts TA (1987) The effect of sodium chloride and temperature on the rate and extent of growth of *Clostridium botulinum* type A in pasteurised pork slurry. J Appl Bacteriol 62:479–490. doi:10.1111/j.1365-2672.1987.tb02680.x

Gibson A, Bratchell N, Roberts T (1988) Predicting microbial growth: growth responses of Salmonellae in a laboratory medium as affected by pH, sodium chloride and storage temperature. Int J Food Microbiol 6:155–178. doi:10.1016/0168-1605(88)90051-7

Gimenez B, Dalgaard P (2004) Modelling and predicting the simultaneous growth of *Listeria monocytogenes* and spoilage micro-organisms in cold-smoked salmon. J Appl Microbiol 96:96–109. doi:10.1046/j.1365-2672.2003.02137.x

Gonzales-Barrón U, Butler F (2011) The use of meta-analytical tools in risk assessment for food safety. Food Microbiol 28:823–827. doi:10.1016/j.fm.2010.04.007

Gorris LGM (2005) Food safety objective: an integral part of food chain management. Food Cont 16:801–809. doi:10.1016/j.foodcont.2004.10.020

Gougouli M, Koutsoumanis KP (2012) Modeling germination of fungal spores at constant and fluctuating temperature conditions. Int J Food Microbiol 152:153–161. doi:10.1016/j.ijfoodmicro.2011.07.030

Guillier L, Pardon P, Augustin JC (2005) Influence of stress on individual lag time distributions of *Listeria monocytogenes*. Appl Environ Microbiol 71:2940–2948. doi:10.1128/AEM.71.6.2940-2948.2005

Guillier L, Pardon P, Augustin JC (2006) Automated image analysis of bacterial colony growth as a tool to study individual lag time distributions of immobilised cells. J Microbiol Methods 65:324–334. doi:10.1016/j.mimet.2005.08.007

Gysemans KPM, Bernaerts K, Vermeulen A, Geeraerd AH, Debevere J, Devlieghere F, Van Impe JF (2007) Exploring the performance of logistic regression model types on growth/no growth data of *Listeria monocytogenes*. Int J Food Microbiol 114:316–331. doi:10.1016/j.ijfoodmicro.2006.09.026

Haas CN, Rose JB, Gerba CP (1999) Quantitative microbial risk assessment. Wiley, New York

Hahne H, Mäder U, Otto A, Bonn F, Steil L, Bremer E, Hecker N, Becher D (2010) A comprehensive proteomics and transcriptomics analysis of *Bacillus subtilis* salt stress adaptation. J Bacteriol 192:870–882. doi:10.1128/JB.01106-09

Hajmeer M, Basheer I, Najjar Y (1997) Computational neural networks for predictive microbiology II. Application to microbial growth. Int J Food Microbiol 34:51–66. doi:10.1016/S0168-1605(96)01169-5

Havelaar AH, Nauta MJ, Jansen JT (2004) Fine-tuning food safety objectives and risk assessment. Int J Food Microbiol 93:11–29. doi:10.1016/j.ijfoodmicro.2003.09.012

Heath AP, Kavraki LE (2009) Computational challenges in systems biology. Comput Sci Rev 3:1–17. doi:10.1016/j.cosrev.2009.01.002

Hertog MLATM, Rudell DR, Pedreschi R, Schaffer RJ, Geeraerd AH, Nicolai BM, Ferguson I (2011) Where systems biology meets postharvest. Postharvest Biol Technol 62:223–237. doi:10.1016/j.postharvbio.2011.05.007

Hervás C, Zurera G, García-Gimeno RM, Martinez J (2001) Optimization of computational neural network for its application to the prediction of microbial growth in foods. Food Sci Technol Int 7:159–163. doi:10.1106/6Q2A-8D7R-JHJU-T7F6

Hervás-Martínez C, García-Gimeno RM, Martínez-Estudillo AC, Martínez-Estudillo FJ, Zurera-Cosano G (2006) Improving microbial growth prediction by Product Unit Neural Networks. J Food Sci 71(2):31–38. doi:10.1111/j.1365-2621.2006.tb08904.x

Hills BP, Mackey BM (1995) Multicompartment kinetic models for injury, resuscitation induced lag and growth in bacterial-cell populations. J Theor Biol 12:333–346. doi:10.1016/S0740-0020(95)80114-6

Hills BP, Manning CE, Ridge YP, Brocklehurst TF (1997) Water availability and the survival of *Salmonella typhimurium* in porous systems. Int J Food Microbiol 36:187–198. doi:10.1016/S0168-1605(97)01265-8

Hoffman FO, Hammonds JS (1994) Propagation of uncertainty in risk assessments: the need to distinguish between uncertainty due to lack of knowledge and uncertainty due to variability. Risk Anal 14:707–712. doi:10.1111/j.1539-6924.1994.tb00281.x

Holm C, Mathiasen T, Jespersen L (2004) A flow cytometric technique for quantification and differentiation of bacteria in bulk tank milk. J Appl Microbiol 97:935–941. doi:10.1111/j.1365-2672.2004.02346.x

Huang L, Hwang A, Phillips J (2011) Effect of temperature on microbial growth rate-mathematical analysis: the Arrhenius and Eyring–Polanyi connections. J Food Sci 76:553–560. doi:10.1111/j.1750-3841.2011.02377.x

ICMSF (International Commission on the Microbiological Specifications of Foods) (2002) Microorganisms in Foods. 7. Microbiological testing in food safety management. Kluwer/ Plenum, New York

ILSI (International Life Sciences Institute). Europe risk analysis in food microbiology task force industry members (2010) Impact of microbial distributions on food safety. ISBN 9789078637202

Ivanek R, Gröhn YT, Wiedmann M, Wells MT (2004) Mathematical model of *Listeria monocytogenes* cross-contamination in a fish processing plant. J Food Prot 67:2688–2697

Jablasone J, Warriner K, Griffiths M (2005) Interactions of *Escherichia coli* O157:H7, *Salmonella typhimurium* and *Listeria monocytogenes* plants cultivated in a gnotobiotic system. Int J Food Microbiol 99:7–18. doi:10.1016/j.ijfoodmicro.2004.06.011

Jameson J (1962) A discussion of the dynamics of *Salmonella* enrichment. J Hyg 60:193–207

Janevska DP, Gospavic R, Pacholewicz E, Popov V (2010) Application of a HACCP–QMRA approach for managing the impact of climate change on food quality and safety. Food Res Int 43:1915–1924. doi:10.1016/j.foodres.2010.01.025

Jasson V, Jacxsens L, Luning P, Rajkovic A, Uyttendaele M (2010) Alternative microbial methods: an overview and selection criteria. Food Microbiol 27:710–730. doi:10.1016/j. fm.2010.04.008

Jeyamkondan S, Jayas DS, Holley RA (2001) Microbial growth modelling with artificial neural networks. Int J Food Microbiol 64:343–354. doi:10.1016/S0168-1605(00)00483-9

Jouve JL (1999) Establishment of food safety objectives. Food Cont 10:303–305. doi:10.1016/ S0956-7135(99)00059-6

Judet-Correia D, Bollaert S, Duquenne A, Charpentier C, Bensoussan M, Dantigny P (2010) Validation of a predictive model for the growth of *Botrytis cinerea* and *Penicillium expansum* on grape berries. Int J Food Microbiol 142:106–113. doi:10.1016/j.ijfoodmicro.2010.06.009

Julien E, Boobis AR, Olin SS (2009) The key events dose–response framework: a cross-disciplinary mode-of-action based approach to examining dose–response and thresholds. Crit Rev Food Sci Nutr 49:682–689. doi:10.1080/10408390903110692

Juneja JK, Marmer BS, Phillips JG, Miller AJ (1995) Influence of the intrinsic properties of food on thermal inactivation of spores of nonproteolytic *Clostridium botulinum*: development of a predictive model. J Food Saf 15:349–364. doi:10.1111/j.1745-4565.1995.tb00145.x

Kahm M, Hasenbrink G, Lichtenberg-Frate H et al (2010) Grofit: fitting biological growth curves with R. J Stat Softw 33:1–21

Karadavut U, Palta Ç, Kökten K, Bakoğlu A (2010) Comparative study on some non-linear growth models for describing leaf growth of maize. Int J Agric Biol 12:227–230

Kauffman KJ, Prakash P, Edwards JS (2003) Advances in flux balance analysis. Curr Opin Biotechnol 14:491–496. doi:10.1016/j.copbio.2003.08.001

Keijser BJF, Ter Beek A, Rauwerda H, Schuren F, Montijn R, van der Spek H, Brul S (2007) Analysis of temporal gene expression during Bacillus subtilis spore germination and outgrowth. J Bacteriol 189:3624–3634

Keskinen L, Todd ECD, Ryser ET (2008) Transfer of surface-dried *Listeria monocytogenes* from stainless steel knife blades to roast turkey breast. J Food Prot 71:176–181

Kitano H (2002) Systems biology: a brief overview. Science 295:1662–1664. doi:10.1126/ science.1069492

Klapwijk PM, Jouve JL, Stringer MF (2000) Microbiological risk assessment in Europe: the next decade. Int J Food Microbiol 58:223–230. doi:10.1016/S0168-1605(00)00276-2

Koseki S (2009) Microbial responses viewer (MRV): a new ComBase-derived database of microbial responses to food environments. Int J Food Microbiol 134:75–82. doi:10.1016/j. ijfoodmicro.2008.12.019

Koutsoumanis K (2001) Predictive modeling of the shelf life of fish under nonisothermal conditions. Appl Environ Microbiol 67:1821–1829. doi:10.1128/AEM.67.4.1821-1829.2001

Koutsoumanis K (2008) A study on the variability in the growth limits of individual cells and its effect on the behavior of microbial populations. Int J Food Microbiol 128:116–121. doi:10.1016/j.ijfoodmicro.2008.07.013

Koutsoumanis K, Nychas GJE (2000) Application of a systematic experimental procedure to develop a microbial model for rapid fish shelf life prediction. Int J Food Microbiol 60:171–184. doi:10.1016/S0168-1605(00)00309-3

Koutsoumanis K, Kendall PA, Sofos J (2004a) A comparative study on growth limits of *Listeria monocytogenes* as affected by temperature, pH and a_w when grown in suspension or on a solid surface. Food Microbiol 21:415–422. doi:10.1016/j.fm.2003.11.003

Koutsoumanis KP, Kendall PA, Sofos JN (2004b) Modeling the boundaries of growth of *Salmonella typhimurium* in broth as a function of temperature, water activity, and pH. J Food Prot 67:53–59

Koutsoumanis K, Taoukis PS, Nychas GJE (2005) Development of a safety monitoring and assurance system for chilled food products. Int J Food Microbiol 100:253–260. doi:10.1016/j.ijfoodmicro.2004.10.024

Koutsoumanis K, Stamatiou A, Skandamis P, Nychas GE (2006) Development of a microbial model for the combined effect of temperature and pH on spoilage of ground meat, and validation of the model under dynamic temperature conditions. Appl Environ Microbiol 72:124–134. doi:10.1128/AEM.72.1.124

Lammerding AM, Fazil A (2000) Hazard identification and exposure assessment for microbial food safety risk assessment. Int J Food Microbiol 58:147–157. doi:10.1016/S0168-605(00)00269-5

Lammerding AM, Paoli GM (1997) Quantitative risk assessment: an emerging tool for emerging foodborne pathogens. Emerg Infect Dis 3:483–487. doi:10.3201/eid0304.970411

Lanciotti R, Sinigaglia M, Gardini F, Vannini L, Guerzoni ME (2001) Growth/no growth interfaces of *Bacillus cereus*, *Staphylococcus aureus* and *Salmonella enteritidis* in model systems based on water activity, pH, temperature and ethanol concentration. Food Microbiol 18:659–668. doi:10.1006/fmic.2001.0429

Lanzanova M, Mucchetti G, Neviani E (1993) Analysis of conductance changes as a growth index of lactic acid bacteria in milk. J Dairy Sci 76:20–28. doi:10.3168/jds.S0022-0302(93)77319-1

Larsen P, Hamada Y, Gilbert J (2012) Modeling microbial communities: current, developing, and future technologies for predicting microbial community interaction. J Biotechnol. doi:10.1016/j.jbiotec.2012.03.009

Le Marc Y, Huchet V, Bourgeois CM, Guyonnet JP, Mafart P, Thuault D (2002) Modelling the growth kinetics of *Listeria* as a function of temperature, pH and organic acid concentration. Int J Food Microbiol 73:219–237. doi:10.1016/S0168-1605(01)00640-7

Le Marc Y, Pin C, Baranyi J (2005) Methods to determine the growth domain in a multidimensional environmental space. Int J Food Microbiol 100:3–12. doi:10.1016/j.ijfoodmicro.2004.10.003

Le Marc Y, Valík L, Medvedová A (2009) Modelling the effect of the starter culture on the growth of *Staphylococcus aureus* in milk. Int J Food Microbiol 129:306–311. doi:10.1016/j.ijfoodmicro.2008.12.015

Lebert I, Robles-Olvera V, Lebert A (2000) Application of polynomial models to predict growth of mixed cultures of *Pseudomonas* spp. and *Listeria* in meat. Int J Food Microbiol 61:27–39. doi:10.1016/S0168-1605(00)00359-7

Lee SH, Hou CL (2002) An art-based construction of rbf networks. IEEE Trans Neural Netw 13 (6):1308–1321

Leguérinel I, Mafart P (1998) Model for combined effects of temperature, pH and water activity on thermal inactivation of *Bacillus cereus* spores. J Food Sci 63:887–889. doi:10.1111/j.1365-2621.1998.tb17920.x

Leistner L (1992) Food preservation by combined methods. Food Res Int 25:151–158. doi:10.1016/0963-9969(92)90158-2

Leporq B, Membré JM, Dervin C, Buche P, Guyonnet JP (2005) The 'Sym'Previus' software, a tool to support decisions to the foodstuff safety. Int J Food Microbiol 100:231–237. doi:10.1016/j.ijfoodmicro.2004.10.006

Leroi F, De Vuyst L (2007) Modelling microbial interactions in foods. In: Brul S, van Gerwen SJC, Zwietering MH (eds) Modelling microorganisms in food. CRC Press, Boca Raton, pp 214–224

Leroi F, Fall PA, Pilet MF et al (2012) Influence of temperature, pH and NaCl concentration on the maximal growth rate of *Brochothrix thermosphacta* and a bioprotective bacteria *Lactococcus piscium* CNCM I-4031. Food Microbiol 31:222–228. doi:10.1016/j.fm.2012.02.014

Lianou A, Koutsoumanis KP (2011) Effect of the growth environment on the strain variability of *Salmonella enterica* kinetic behavior. Food Microbiol 28:828–837. doi:10.1016/j.fm.2010.04.006

Lindberg CW, Borch E (1994) Predicting the aerobic growth of *Yersinia enterocolitica* O:3 at different pH-values, temperatures and L-lactate concentrations using conductance measurements. Int J Food Microbiol 22:141–153. doi:10.1016/0168-1605(94)90138-4

Lindqvist R (2006) Estimation of *Staphylococcus aureus* growth parameters from turbidity data: characterization of strain variation and comparison of methods. Appl Environ Microbiol 72:4862–4870. doi:10.1128/AEM.00251-06

Lindqvist R, Sylvén S, Vågsholm I (2002) Quantitative microbial risk assessment exemplified by *Staphylococcus aureus* in unripened cheese made from raw milk. Int J Food Microbiol 78:155–170. doi:10.1016/S0168-1605(02)00237-4

Mafart P, Leguérinel I (1998) Modeling combined effects of temperature and pH on heat resistance of spores by a linear-Bigelow equation. J Food Sci 63:6–8. doi:10.1111/j.1365-2621.1998.tb15662.x

Malakar PK, Baker GC (2008) Estimating single-cell lag times via a Bayesian scheme. Appl Environ Microbiol 74:7098–7099. doi:10.1128/AEM.01277-08

Malakar P, Barker G, Zwietering M, van't Riet K (2003) Relevance of microbial interactions to predictive microbiology. Int J Food Microbiol 84:263–272. doi:10.1016/S0168-1605(02)00424-5

Martens DE, Béal C, Malakar P et al (1999) Modelling the interactions between *Lactobacillus curvatus* and *Enterobacter cloacae*. I. Individual growth kinetics. Int J Food Microbiol 51:53–65. doi:10.1016/S0168-1605(99)00095-1

Masana MO, Baranyi J (2000) Adding new factors to predictive models: the effect on the risk of extrapolation. Food Microbiol 17:367–374. doi:10.1006/fmic.1999.0326

Mataragas M, Drosinos EH, Vaidanis A, Metaxopoulos I (2006) Development of a predictive model for spoilage of cooked cured meat products and its validation under constant and dynamic temperature storage conditions. J Food Sci 71:M157–M167. doi:10.1111/j.1750-3841.2006.00058.x

Mataragas M, Zwietering MH, Skandamis PN, Drosinos EH (2010) Quantitative microbiological risk assessment as a tool to obtain useful information for risk managers–specific application to *Listeria monocytogenes* and ready-to-eat meat products. Int J Food Microbiol 141((suppl)): S170–S179. doi:10.1016/j.ijfoodmicro.2010.01.005

McAdams H, Arkin A (1997) Stochastic mechanisms in gene expression. Proc Natl Acad Sci USA 94:814–819. doi:10.1073/pnas.94.3.814

McBain AJ (2009) In vitro biofilm models: an overview. Adv Appl Microbiol, pp 69:99–132. DOI: 10.1016/S0065-2164(09)69004-3

McClure PJ, Baranyi J, Boogard E, Kelly TM, Roberts TA (1993) A predictive model for the combined effect of pH, sodium chloride and storage temperature on the growth of *Brochothrix thermosphacta*. Int J Food Microbiol 19:161–178. doi:10.1016/0168-1605(93)90074-Q

McClure PJ, de Blackburn CW, Cole MB, Curtis PS, Jones JE, Legan JD, Ogden ID, Peck MW, Roberts TA, Sutherland JP, Walker SJ (1994) Modelling the growth, survival and death of microorganisms in foods: the UK Food MicroModel approach. Int J Food Microbiol 23. doi:10.1016/0168-1605(94)90156-2

McClure PJ, Beaumont AL, Sutherland JP, Roberts TA (1997) Predictive modelling of growth of *Listeria monocytogenes*. The effects on growth of NaCl, pH, storage temperature and NaNO. Int J Food Microbiol 34:221–232. doi:10.1016/S0168-1605(96)01193-2

McDonald K, Sun DW (1999) Predictive food microbiology for the meat industry: a review. Int J Food Microbiol 52:1–27. doi:10.1016/S0168-1605(99)00126-9

McKellar RC (2001) Development of a dynamic continuous-discrete-continuous model describing the lag phase of individual bacterial cells. J Appl Microbiol 90:407–413. doi:10.1046/j.1365-2672.2001.01258.x

McKellar RC, Knight KP (2000) A combined discrete-continuous model describing the lag phase of *Listeria monocytogenes*. Int J Food Microbiol 54:171–180. doi:10.1016/S0168-1605(99)00204-4

McKellar RC, Lu XW (2001) A probability of growth model for *Escherichia coli* O157:H7 as a function of temperature, pH, acetic acid, and salt. J Food Prot 64:1922–1928

McKellar RC, Lu X (2004) Modelling microbial responses in food, CRC Series in Contemporary Food Science. CRC, London. ISBN 0-8493-1237-X

McKellar RC, Butler G, Stanich K (1997) Modelling the influence of temperature on the recovery of *Listeria monocytogenes* from heat injury. Food Microbiol 14:617–625. doi:10.1006/fmic.1997.0124

McKellar R, Lu X, Delaquis P (2002) A probability model describing the interface between survival and death of *Escherichia coli* O157:H7 in a mayonnaise model system. Food Microbiol 19:235–247. doi:10.1006/fmic.2001.0449

McMeekin TA, Ross T (2002) Predictive microbiology: providing a knowledge-based framework for change management. Int J Food Microbiol 78:133–153. doi:10.1016/S0168-1605(02)00231-3

McMeekin TA, Olley J, Ross T, Ratkowsky DA (1993a) Predictive microbiology: theory and application. Research Studies Press, Taunton

McMeekin TA, Olley JN, Ross T, Ratkowsky DA (1993b) Predictive microbiology: theory and application. Trends Food Sci Technol 4:340. doi:10.1016/0924-2244(93)90049-G

McMeekin TA, Presser KA, Ratkowsky DA, Ross T, Salter M, Tienungoon S (2000) Quantifying the hurdle concept by modelling the bacterial growth/no growth interface. Int J Food Microbiol 55:93–98. doi:10.1016/S0168-1605(00)00182-3

McMeekin TA, Olley J, Ratkowsky DA, Ross T (2002) Predictive microbiology: towards the interface and beyond. Int J Food Microbiol 73:395–407. doi:10.1016/S0168-1605(01)00663-8

McMeekin TA, Baranyi J, Bowman J, Dalgaard P, Kirk M, Ross T, Schmid S, Zwietering MH (2006) Information systems in food safety management. Int J Food Microbiol 112:181–194. doi:10.1016/j.ijfoodmicro.2006.04.048

Mejlholm O, Dalgaard P (2007) Modeling and predicting the growth of lactic acid bacteria in lightly preserved seafood and their inhibiting effect on *Listeria monocytogenes*. J Food Prot 70:2485–2497

Mejlholm O, Gunvig A, Borggaard C, Blom-Hanssen J, Mellefont L, Ross T, Leroi F, Else T, Visser D, Dalgaard P (2010) Predicting growth rates and growth boundary of *Listeria monocytogenes*. An international validation study with focus on processed and ready-to-eat meat and seafood. Int J Food Microbiol 141:137–150. doi:10.1016/j.ijfoodmicro.2010.04.026

Membré JM, Lambert R (2008) Application of predictive modelling techniques in industry: from food design up to risk assessment. Int J Food Microbiol 128:10–15. doi:10.1016/j.ijfoodmicro.2008.07.006

Membré JM, Ross T, McMeekin TA (1999) Behaviour of *Listeria monocytogenes* under combined chilling processes. Lett Appl Microbiol 28:216–220. doi:10.1046/j.1365-2672.1999.00499.x

Membré JM, Leporq B, Vialette M, Mettler E, Perrier L, Thuault D, Zwietering M (2005) Temperature effect on bacterial growth rate: quantitative microbiology approach including cardinal values and variability estimates to perform growth simulations on/in food. Int J Food Microbiol 100:179–186. doi:10.1016/j.ijfoodmicro.2004.10.015

Mertens L, Van Derlinden E, Van Impe JF (2012) Comparing experimental design schemes in predictive food microbiology: optimal parameter estimation of secondary models. J Food Eng. doi:10.1016/j.jfoodeng.2012.03.018

Métris A, George SM, Peck MW, Baranyi J (2002) Effect of sodium chloride and pH on the distribution of the lag times of individual cells of *Listeria innocua*. In: Duby C, Cassar JP (eds) Proceedings of the 7th European conference food industry and statistics. Lille-Cit Scientifique, Villeneuve d'Asq, France, pp 61–66. ISBN 2-7380-1016-4

Métris A, George SM, Peck MW, Baranyi J (2003) Distribution of turbidity detection times produced by single cell-generated bacterial populations. J Microbiol Methods 55:821–827. doi:10.1016/j.mimet.2003.08.006

Métris A, Le Marc Y, Elfwing A, Ballagi A, Baranyi J (2005) Modelling the variability of lag times and the first generation times of single cells of *E. coli*. Int J Food Microbiol 100:13–19. doi:10.1016/j.ijfoodmicro.2004.10.004

Métris A, George SM, Mackey BM, Baranyi J (2008) Modeling the variability of single-cell lag times for *Listeria innocua* populations after sublethal and lethal heat treatments. Appl Environ Microbiol 74:6949–6955. doi:10.1128/AEM.01237-08

Métris A, George S, Baranyi J (2011) Modelling osmotic stress by flux balance analysis at the genomic scale. Int J Food Microbiol 152:123–128. doi:10.1016/j.ijfoodmicro.2011.06.016

Miller FA, Ramos B, Gil MM, Brandao TRS, Teixeira P, Silva CLM (2009) Influence of pH, type of acid and recovery media on the thermal inactivation of *Listeria innocua*. Int J Food Microbiol 133:121–128. doi:10.1016/j.ijfoodmicro.2009.05.007

Møller CO, Nauta MJ, Christensen BB, Dalgaard P, Hansen TB (2012) Modelling transfer of *Salmonella typhimurium* DT104 during simulation of grinding of pork. J Appl Microbiol 112:90–98. doi:10.1111/j.1365-2672.2011.05177.x

Monod J (1949) The growth of bacterial cultures. Annu Rev Microbiol 3:371–394. doi:10.1146/annurev.mi.03.100149.002103

Montville R, Schaffner DW (2003) Inoculum size influences bacterial cross contamination between surfaces. Appl Environ Microbiol 69:7188–7193. doi:10.1128/AEM.69.12.7188

Montville R, Chen Y, Schaffner DW (2001) Glove barriers to bacterial cross-contamination between hands to food. J Food Prot 64:845–849

Mossel DAA (1995) Essentials of the microbiology of foods: a textbook for advanced studies (with the collaboration of Anthony C. Baird-Parker). Wiley, New York

Mylius SD, Nauta MJ, Havelaar AH (2007) Cross-contamination during food preparation: a mechanistic model applied to chicken-borne *Campylobacter*. Risk Anal 27:803–813. doi:10.1111/j.1539-6924.2006.00872.x

National Research Council (NRC) (1994) Science and judgment in risk assessment. National Academy Press, Washington, DC

Nauta MJ (2000) Separation of uncertainty and variability in quantitative microbial risk assessment models. Int J Food Microbiol 57:9–18. doi:10.1016/S0168-1605(00)00225-7

Nauta MJ (2001) Modular process risk model structure for quantitative microbiological risk assessment and its application in an exposure assessment of *Bacillus cereus* in a REPFED. National Institute of Public Health and the Environment, Bilthoven

Nauta M (2003) A retail and consumer phase model for exposure assessment of *Bacillus cereus*. Int J Food Microbiol 83:205–218. doi:10.1016/S0168-1605(02)00374-4

Nauta MJ (2005) Microbiological risk assessment models for partitioning and mixing during food handling. Int J Food Microbiol 100:311–322. doi:10.1016/j.ijfoodmicro.2004.10.027

Nauta MJ (2007) Uncertainty and variability predictive models of microorganisms in food. In: Brul S, van Gerwen SJ, Zwietering MH (eds) Modelling microorganisms in food. CRC Press, Boca Raton, pp 44–65

Nerbrink E, Borch E, Blom H, Nesbakken T (1999) A model based on absorbance data on the growth rate *of Listeria monocytogenes* and including the effects of pH, NaCl, Na-lactate and Na-acetate. Int J Food Microbiol 47:99–109. doi:10.1016/S0168-1605(99)00021-5

Niven GW, Fuks T, Morton JS, Rua SACG, Mackey BM (2006) A novel method for measuring lag times in division of individual bacterial cells using image analysis. J Microbiol Methods 65:311–317. doi:10.1016/j.mimet.2005.08.006

Nixon PA (1971) Temperature integration as a means of assessing storage conditions. Report on quality in fish products. Seminar No. 3. Fishing Industry Board, New Zealand, pp 33–44

Noble PA (1999) Hypothetical model for monitoring microbial growth by using capacitance measurements: a mini review. J Microbiol Methods 37:45–49

Noriega E, Laca A, Diaz M (2008) Modelling of diffusion-limited growth to predict *Listeria* distribution in structured model foods. J Food Eng 87:247–256. doi:10.1016/j.jfoodeng. 2007.11.035

Ongeng D, Ryckeboer J, Vermeulen A, Devlieghere F (2007) The effect of micro-architectural structure of cabbage substratum and or background bacterial flora on the growth of *Listeria monocytogenes*. Int J Food Microbiol 119:291–299. doi:10.1016/j.ijfoodmicro.2007.08.022

Oscar TP (2009) General regression neural network and Monte Carlo simulation model for survival and growth of *Salmonella* on raw chicken skin as a function of serotype, temperature, and time for use in risk assessment. J Food Prot 72:2078–2087

Oscar TP (2011) Plenary lecture: innovative modeling approaches applicable to risk assessments. Food Microbiol 28:777–781. doi:10.1016/j.fm.2010.05.017

Ottoson JR, Nyberg K, Lindqvist R, Albihn A (2011) Quantitative microbial risk assessment for *Escherichia coli* O157 on lettuce, based on survival data from controlled studies in a climate chamber. J Food Prot 74:2000–2007. doi:10.4315/0362-028X.JFP-10-563

Peck MW, Stringer SC, Carter AT (2011) *Clostridium botulinum* in the post-genomic era. Food Microbiol 28:183–191. doi:10.1016/j.fm.2010.03.005

Pérez-Rodríguez F, Todd ECD, Valero A, Carrasco E, García RM, Zurera G (2006) Linking quantitative exposure assessment and risk management using the food safety objective concept: an example with *Listeria monocytogenes* in different cross-contamination scenarios. J Food Prot 69:2384–2394

Pérez-Rodríguez F, Valero A, Todd E, Carrasco E, García-Gimeno RM, Zurera G (2007a) Modeling transfer of *Escherichia coli* O157:H7 and *Staphylococcus aureus* during slicing of a cooked meat product. Meat Sci 76:692–699. doi:10.1016/j.meatsci.2007.02.011

Pérez-Rodríguez F, van Asselt ED, Garcia-Gimeno RM, Zurera G, Zwietering MH (2007b) Extracting additional risk managers information from a risk assessment of *Listeria monocytogenes* in deli meats. J Food Prot 70:1137–1152

Pérez-Rodríguez F, Valero A, Carrasco E, García-Gimeno RM, Zurera G (2008) Understanding and modelling bacterial transfer to foods: a review. Trends Food Sci Technol 19:131–144. doi:10.1016/j.tifs.2007.08.003

Pérez-Rodríguez F, Campos D, Ryser ET, Buchholz AL, Posada-Izquierdo GD, Marks BP, Zurera G, Todd ECD (2011) A mathematical risk model for *Escherichia coli* O157:H7 cross-contamination of lettuce during processing. Food Microbiol 28:694–701. doi:10.1016/j. fm.2010.06.008

Peterson AC, Black JJ, Gunderson MF (1964) Staphylococci in competition. III. Influence of pH and salt on staphylococcal growth in mixed populations. Appl Microbiol 12:70–76

Petranovic D, Vemuri GN (2009) Impact of yeast systems biology on industrial biotechnology. J Biotechnol 144:204–211. doi:10.1016/j.jbiotec.2009.07.005

Pin C, Sutherland JP, Baranyi J (1999) Validating predictive models of food spoilage organisms. J Appl Microbiol 87:491–499. doi:10.1046/j.1365-2672.1999.00838.x

Pin C, Baranyi J, de Fernando GG (2000) Predictive model for the growth of *Yersinia enterocolitica* under modified atmospheres. J Appl Microbiol 88:521–530. doi:10.1046/ j.1365-2672.2000.00991.x

Pin C, Avendaño-Pérez G, Cosciani E, Gómez N, Gounadakic A, Nychas G, Skandamis P, Barker G (2011) Modelling *Salmonella* concentration throughout the pork supply chain by considering growth and survival in fluctuating conditions of temperature, pH and a_w. Int J Food Microbiol 145:S96–S102. doi:0.1016/j.ijfoodmicro.2010.09.025

Polese P, Del Torre M, Spaziani M, Stecchini ML (2011) A simplified approach for modelling the bacterial growth/no growth boundary. Food Microbiol 28:384–391. doi:10.1016/j. fm.2010.09.011

Poschet F (2003) Monte Carlo analysis as a tool to incorporate variation on experimental data in predictive microbiology. Food Microbiol 20:285–295. doi:10.1016/S0740-0020(02)00156-9

Poschet F, Van Impe JF (1999) Quantifying the uncertainty of model outputs in predictive microbiology: a Monte-Carlo analysis. Meded Fac Landbouwwet Univ Gent 64(5):499–506

Poschet F, Vereecken KM, Geeraerd AH et al (2005) Analysis of a novel class of predictive microbial growth models and application to coculture growth. Int J Food Microbiol 100:107–124. doi:10.1016/j.ijfoodmicro.2004.10.008

Pouillot R, Delignette-Muller ML (2010) Evaluating variability and uncertainty separately in microbial quantitative risk assessment using two R packages. Int J Food Microbiol 142:330–340. doi:10.1016/j.ijfoodmicro.2010.07.011

Pouillot R, Albert I, Cornu M, Denis JB (2003) Estimation of uncertainty and variability in bacterial growth using Bayesian inference. Application to *Listeria monocytogenes*. Int J Food Microbiol 81:87–104. doi:10.1016/S0168-1605(02)00192-7

Pouillot R, Miconnet N, Afchain AL, Delignette-Müller ML, Beaufort A, Rosso L, Denis JB, Cornu M (2007) Quantitative risk assessment of *Listeria monocytogenes* in French cold-smoked salmon: I. Quantitative exposure assessment. Risk Anal 27:683–700. doi:10.1111/j.1539-6924.2007.00921.x

Powell M (2004) Considering the complexity of microbial community dynamics in food safety risk assessment. Int J Food Microbiol 90:171–179. doi:10.1016/S0168-1605(03)00106-5

Presser KA, Ratkowsky DA, Ross T (1997) Modelling the growth rate of *Escherichia coli* as a function of pH and lactic acid concentration. Appl Environ Microbiol 63:2355–2360

Presser KA, Ross T, Ratkowsky DA (1998) Modeling the growth limits (growth/no growth interface) of *Escherichia coli* as a function of temperature, pH, lactic acid concentration, and water activity. Appl Environ Microbiol 64:1773–1779

Psomas AN, Nychas GJ, Haroutounian SA, Skandamis PN (2011) Development and validation of a tertiary simulation model for predicting the growth of the food microorganisms under dynamic and static temperature conditions. Comput Electron Agric 76:119–129. doi:10.1016/j.compag.2011.01.013

Psomas AN, Nychas GJ, Haroutounian SA, Skandamis PN (2012) LabBase: development and validation of an innovative food microbial growth responses database. Comput Electron Agric 85:99–108. doi:10.1016/j.compag.2012.04.002

Rasch M (2004) Experimental design and data collection. In: Mckellar RC, Lu X (eds) Modelling microbial response in food. CRC Press, Boca Raton, pp 1–20

Rasch M, Métris A, Baranyi J, Budde BB (2007) The effect of reuterin on the lag time of single cells of *Listeria innocua* grown on a solid agar surface at different pH and NaCl concentrations. Int J Food Microbiol 113:35–40. doi:10.1016/j.ijfoodmicro.2006.07.012

Ratkowsky DA (ed) (1983) Nonlinear regression modeling: a unified practical approach. Dekker, New York

Ratkowsky DA (2002) Some examples of, and some problems with, the use of nonlinear logistic regression in predictive food microbiology. Int J Food Microbiol 73:119–125. doi:10.1016/S0168-1605(01)00643-2

Ratkowsky DA (2004) Model fitting and uncertainty. In: McKellar RC, Lu X (eds) Modelling microbial responses in foods. CRC Press, Boca Raton, pp 191–195

Ratkowsky DA, Ross T (1995) Modelling the bacterial growth/no growth interface. Lett Appl Microbiol 20:29–33. doi:10.1111/j.1472-765X.1995.tb00400.x

Ratkowsky DA, Olley J, McMeekin TA, Ball A (1982) Relationship between temperature and growth rates of bacterial cultures. J Bacteriol 149:1–5

Ratkowsky DA, Lowry RK, McMeekin TA, Stokes AN, Chandler RE (1983) Model for bacterial culture growth rate throughout the entire biokinetic temperature range. J Bacteriol 154:1222–1226

Reed JL, Palsson BØ (2003) Thirteen years of building constraint-based in silico models of *Escherichia coli*. J Bacteriol Soc. doi:10.1128/JB.185.9.2692

Reichart O (1994) Modeling the destruction of *Escherichia coli* on the base of reaction kinetics. Int J Food Microbiol 23:449–465. doi:10.1016/0168-1605(94)90169-4

Roberts D (1990) Foodborne illness; sources of infection: food. Lancet 336:859–861. doi:10.1016/0140-6736(90)92352-I

Roberts TA, Jarvis B (1983) Predictive modelling of food safety with particular reference to *Clostridium botulinum* in model cured meat systems. In: Roberts TA, Skinner FA (eds) Food microbiology: advances and prospects. Academic Press, New York, pp 85–95

Roberts TA, Gibson AM, Robinson A (1981) Prediction of toxin production by *Clostridium botulinum* in pasteurised pork slurry. J Food Technol 16:337–355

Robinson TP, Ocio MJ, Kaloti A, Mackey BM (1998) The effect of the growth environment on the lag phase of *Listeria monocytogenes*. Int J Food Microbiol 44:83–92. doi:10.1016/S0168-1605(98)00120-2

Robinson TP, Aboaba OO, Kaloti A, Ocio MJ, Baranyi J, Mackey BM (2001) The effect of inoculum size on the lag phase of *Listeria monocytogenes*. Int J Food Microbiol 70:163–173. doi:10.1016/S0168-1605(01)00541-4

Ross T (1996) Indice of performance evaluation of predictive models in food microbiology. J Appl Bacteriol 81:501–508. doi:10.1111/j.1365-2672.1996.tb03539.x

Ross T, Dalgaard P (2004) Secondary models. In: McKellar RC, Lu X (eds) Modelling microbial responses in foods. CRC Press, Boca Raton, pp 63–150. ISBN 0-8493-1237-X

Ross T, McMeekin TA (1994) Predictive microbiology. Int J Food Microbiol 23:241–264. doi:10.1016/0168-1605(94)90155-4

Ross T, Dalgaard P, Tienungoon S (2000) Predictive modelling of the growth and survival of *Listeria* in fishery products. Int J Food Microbiol 62:231–245. doi:10.1016/S0168-1605(00)00340-8

Ross T, Ratkowsky DA, Mellefont LA, McMeekin TA (2003) Modelling the effects of temperature, water activity, pH and lactic acid concentration on the growth rate of *Escherichia coli*. Int J Food Microbiol 82:33–43. doi:10.1016/S0168-1605(02)00252-0

Ross T, Zhang D, Mc Questin OJ (2008) Temperature governs the inactivation rate of vegetative bacteria under growth-preventing conditions. Int J Food Microbiol 128:129–135. doi:10.1016/j.ijfoodmicro.2008.07.023

Rosset P, Cornu M, Noel V, Morelli E, Poumeyrol G (2004) Time-temperature profiles of chilled ready-to-eat foods in school catering and probabilistic analysis of *Listeria monocytogenes* growth. Int J Food Microbiol 96:49–59. doi:10.1016/j.ijfoodmicro.2004.03.008

Rosso L (1995) Modelling and predictive microbiology: building of a new tool for food industry. Ph.D. thesis, Université Claude Bernard, Lyon, France

Rosso L, Robinson TP (2001) A cardinal model to describe the effect of water activity on the growth of moulds. Int J Food Microbiol 63:265–273. doi:10.1016/S0168-1605(00)00469-4

Rosso L, Lobry JR, Bajard S, Flandrois JP (1995) Convenient model to describe the combined effects of temperature and pH on microbial growth. Appl Environ Microbiol 61:610–616

Rosso L, Bajard S, Flandrois JP, Lahellec C, Fournaud J, Veit P (1996) Differential growth of *Listeria monocytogenes* at 4° and 8°C: consequences for the shelf life of chilled products. J Food Prot 59:944–949

Salter MA, Ross T, Ratkowsky DA, McMeekin TA (2000) Modelling the combined temperature and salt (NaCl) limits for growth of a pathogenic *Escherichia coli* strain using generalised non-linear regression. Int J Food Microbiol 61:159–167. doi:10.1016/S0168-1605(00)00352-4

Schellekens M, Martens T, Roberts TA, Mackey BM, Nicolai BM, Van Impe JF, Debaerdemaeker J (1994) Computer-aided microbial safety design of food processes. Int J Food Microbiol 24:1–9. doi:10.1016/0168-1605(94)90102-3

Schepers A, Thibault J, Lacroix C (2000) Comparison of simple neural networks and nonlinear regression models for descriptive modeling of *Lactobacillus helveticus* growth in pH-controlled batch cultures. Enzyme Microb Technol 26:431–445. doi:10.1016/S0141-0229(99)00183-0

Schuster S, Klamt S, Weckwerth W, Moldenhauer F, Pfieffer T (2002) Use of network analysis of metabolic systems in bioengineering. Bioprocess Biosyst Eng 24:363–372. doi:10.1016/S0167-7799(02)02026-7

Scott WJ (1937) The growth of microorganisms on ox muscle. I. The influence of temperature. J Counc Sci Ind Res Aust 10:338–350

Shadbolt C, Ross T, McMeekin TA (2001) Differentiation of the effects of lethal pH and water activity: food safety implications. Lett Appl Microbiol 32:99–102. doi:10.1046/j.1472-765x.2001.00862.x

Sheen S (2008) Modeling surface transfer of *Listeria monocytogenes* on salami during slicing. J Food Sci 73:E304–E311. doi:10.1111/j.1750-3841.2008.00833.x

Sheen S, Hwang CA (2010) Mathematical modeling the cross-contamination of *Escherichia coli* O157:H7 on the surface of ready-to-eat meat product while slicing. Food Microbiol 27:37–43. doi:10.1016/j.fm.2009.07.016

Shimoni E, Labuza PT (2000) Modelling pathogen growth in meat products: future challenges. Trends Food Sci Technol 11:394–402. doi:10.1016/S0924-2244(01)00023-1

Silva AR, Sant'Ana AS, Massaguer PR (2010) Modelling the lag time and growth rate of *Aspergillus* section *Nigri* IOC 4573 in mango nectar as a function of temperature and pH. J Appl Microbiol 109:1105–1116. doi:10.1111/j.1365-2672.2010.04803.x

Skandamis PN, Stopforth JD, Kendall P, Belk KE, Scanga JA, Smith GC, Sofos JN (2007) Modeling the effect of inoculum size and acid adaptation on growth/no growth interface of *Escherichia coli* O157:H7. Int J Food Microbiol 120:237–249. doi:10.1016/j.ijfoodmicro. 2007.08.028

Smelt JPPM, Otten GD, Bos AP (2002) Modelling the effect of sublethal injury on the distribution of the lag times of individual cells of *Lactobacillus plantarum*. Int J Food Microbiol 73:207–212. doi:10.1016/S0168-1605(01)00651-1

Smyth GK, El-shaarawi AH, Piegorsch WW (2002) Nonlinear regression. Environmetrics 3:1405–1411

Sørensen B, Jakobsen M (1996) The combined effects of temperature, pH and NaCl on growth of *Debaryomyces hansenii* analyzed by flow cytometry and predictive microbiology. Int J Food Microbiol 34:209–220. doi:10.1006/fmic.1996.0032

Spencer R, Baines CR (1964) The effect of temperature on the spoilage of wet fish: I. Storage at constant temperature between -1°C and 25°C. Food Technol Champaign 18:769–772

Stelling J (2004) Mathematical models in microbial systems biology. Curr Opin Microbiol 7:513–518. doi:10.1016/j.mib.2004.08.004

Stewart CM, Cole MB, Legan JD, Slade L, Vandeven MH, Schaffner DW (2001) Modeling the growth boundary of *Staphylococcus aureus* for risk assessment purposes. J Food Prot 64:51–57

Strachan NJC, Doyle MP, Kasuga F, Rotariu O, Ogden ID (2005) Dose response modelling of *Escherichia coli* O157 incorporating data from foodborne and environmental outbreaks. Int J Food Microbiol 103:35–47. doi:10.1016/j.ijfoodmicro.2004.11.023

Stringer M (2005) Summary report. Food safety objectives—role in microbiological food safety management. Food Cont 16:775–794. doi:10.1016/j.foodcont.2004.10.018

Stringer M, George SM, Peck MW (2000) Thermal inactivation of *Escherichia coli* O157:H7. Symp Ser Soc Appl Microbiol 29:79S–89S

Stringer SC, Webb MD, Peck MW (2011) Lag time variability in individual spores of *Clostridium botulinum*. Food Microbiol 28:228–235. doi:10.1016/j.fm.2010.03.003

Stumbo CR, Purohit KS, Ramakrishnan TV (1975) Thermal process lethality guide for low acid foods in metal containers. J Food Sci 40:1316–1323. doi:10.1111/j.1365-2621.1975.tb01080.x

Sumner J, Krist K (2002) The use of predictive microbiology by the Australian meat industry. Int J Food Microbiol 73:363–366. doi:10.1016/S0168-1605(01)00672-9

Sutherland JP, Bayliss AJ (1994) Predictive modelling of growth of *Yersinia enterocolitica*: the effects of temperature, pH and sodium chloride. Int J Food Microbial 21:197–215. doi:10.1016/0168-1605(94)00082-H

Sutherland JP, Bayliss AJ, Roberts TA (1994) Predictive modelling of growth of *Staphylococcus aureus*: the effects of temperature, pH and sodium chloride. Int J Food Microbiol 21:217–236. doi:10.1016/0168-1605(94)90029-9

Tassou CC, Natskoulis PI, Magan N, Panagou EZ (2009) Effect of temperature and water activity on growth and ochratoxin A production boundaries of two *Aspergillus carbonarius* isolates on a simulated grape juice medium. J Appl Microbiol 107:257–268. doi:10.1111/j.1365-2672.2009.04203.x

te Giffel MC, Zwietering MH (1999) Validation of predictive models describing the growth of *Listeria monocytogenes*. Int J Food Microbiol 46:135–149. doi:10.1016/S0168-1605(98)00189-5

Tienungoon S, Ratkowsky DA, McMeekin TA (2000) Growth limits of *Listeria monocytogenes* as a function of temperature, pH, NaCl, and lactic acid. Appl Environ Microbiol 66:4979–4987. doi:10.1128/AEM.66.11.4979-4987.2000

Todd ECD (2004) Microbiological safety standards and public health goals to reduce foodborne disease. Meat Sci 66:33–43. doi:10.1016/S0309-1740(03)00023-8

Toyofoku H (2006) WHO Guides. Guiding future action for food safety: implementation of risk management decision. www.wpro.who.int/fsi_guide/files/implementation_of_risk_management_decision.pps. Accessed 20 Mar 2012

Treviño Santa Cruz MB, Genoud D, Métraux JP, Genoud T (2005) Update in bioinformatics. Toward a digital database of plant cell signalling networks: advantages, limitations and predictive aspects of the digital model. Phytochemistry 66:267–276. doi:10.1016/j.phytochem.2004.11.020

Tromp SO, Rijgersberg H, Franz E (2010) Quantitative microbial risk assessment for *Escherichia coli* O157:H7, *Salmonella enterica*, and *Listeria monocytogenes* in leafy green vegetables consumed at salad bars, based on modeling supply chain logistics. J Food Prot 73:1830–1840

Vaikousi H, Biliaderis CG, Koutsoumanis K (2009) Applicability of a microbial time temperature indicator (TTI) for monitoring spoilage of modified atmosphere packed minced meat. Int J Food Microbiol 133:272–278. doi:10.1016/j.ijfoodmicro.2009.05.030

Valero A, Pérez-Rodríguez F, Carrasco E, García-Gimeno RM, Zurera G (2006) Modeling the growth rate of *Listeria monocytogenes* using absorbance measurements and calibration curves. J Food Sci 71:M257–M264. doi:oi.org/10.1111/j.1750-3841.2006.00139.x

Valero A, Hervás C, García-Gimeno RM, Zurera G (2007) Product unit neural network models for predicting the growth limits of *Listeria monocytogenes*. Food Microbiol 24:452–464. doi:10.1016/j.fm.2006.10.002

Valero A, Pérez-Rodríguez F, Carrasco E, Fuentes-Alventosa JM, Garcia-Gimeno RM, Zurera G (2009) Modelling the growth boundaries of *Staphylococcus aureus*: effect of temperature, pH and water activity. Int J Food Microbiol 133:186–194. doi:10.1016/j.ijfoodmicro.2009.05.023

Valero A, Rodríguez M, Carrasco E, Pérez-Rodríguez F, García-Gimeno RM, Zurera G (2010) Studying the growth boundary and subsequent time to growth of pathogenic *Escherichia coli* serotypes by turbidity measurements. Food Microbiol 27:819–828. doi:10.1016/j.fm.2010.04.016

Van Asselt E, Zwietering MH (2006) A systematic approach to determine global thermal inactivation parameters for various food pathogens. Int J Food Microbiol 107:73–82. doi:10.1016/j.ijfoodmicro.2005.08.014

Van Asselt ED, de Jong EI, de Jonge R, Nauta MJ (2008) Cross-contamination in the kitchen: estimation of transfer rates for cutting boards, hands and knives. J Appl Microbiol 105:1392–1401. doi:10.1111/j.1365-2672.2008.03875.x

Van Besten HMW, Zwietering MH (2012) Meta-analysis for quantitative microbiological risk assessments and benchmarking data. Trends Food Sci Technol 25:34–39. doi:10.1016/j.tifs.2011.12.004

Van Boekel MAJS (2002) On the use of the Weibull model to describe thermal inactivation of microbial vegetative cells. Int J Food Microbiol 74:139–159. doi:10.1016/S0168-1605(01)00742-5

Van Boekel MAJS (2008) Kinetic modelling of food quality: a critical review. Compr Rev Food Sci Food Saf 7:144–158. doi:10.1111/j.1541-4337.2007.00036.x

Van Gerwen SJ, Gorris L (2004) Application of elements of microbiological risk assessment in the food industry via a tiered approach. J Food Prot 67:2033–2040

Van Gerwen SJC, Te Giffel MC, Van't Riet K, Beumer RR, Zwietering MH (2000) Stepwise quantitative risk assessment as a tool for characterization of microbiological food safety. J Appl Microbiol 88:938–951. doi:10.1046/j.1365-2672.2000.01059.x

Van Impe JF, Vercammen D, Van Derlinden E (2011) Developing next generation predictive models: a systems biology approach. Proc Food Sci 1:965–971. doi:10.1016/j.profoo.2011.09.145

Van Schothorst M (2002) Microbiological risk assessment in food processing. In: Implementing the results of a microbiological risk assessment: pathogens risk management. Woodhead, Cambridge, p 188

Van Schothorst MV (2004) A proposed framework for the use of FSOs. Food Cont. doi:10.1016/j.foodcont.2004.10.021

Van Schothorst M (2005) A proposed framework for the use of FSOs. Food Cont 16:811–816. doi:10.1016/j.foodcont.2004.10.021

Varma A, Palsson BO (1994) Metabolic flux balancing: basic concepts, scientific and practical use. Biotechnology 12:994–998. doi:10.1038/nbt1094-994

Vereecken KM, Van Impe JF (2002) Analysis and practical implementation of a model for combined growth and metabolite production of lactic acid bacteria. Int J Food Microbiol 73:239–250. doi:10.1016/S0168-1605(01)00641-9

Vereecken K, Dens EJ, Van Impe J (2000) Predictive modeling of mixed microbial populations in food products: evaluation of two-species models. J Theor Biol 205:53–72

Vereecken KM, Devlieghere F, Bockstaele A et al (2003) A model for lactic acid-induced inhibition of *Yersinia enterocolitica* in mono- and coculture with *Lactobacillus sakei*. Food Microbiol 20:701–713. doi:10.1016/S0740-0020(03)00031-5

Vermeiren L, Devlieghere F, Vandekinderen I, Debevere J (2006) The interaction of the non-bacteriocinogenic *Lactobacillus sakei* 10A and lactocin S producing *Lactobacillus sakei* 148 towards *Listeria monocytogenes* on a model cooked ham. Food Microbiol 23:511–518. doi:10.1016/j.fm.2005.10.005

Vermeulen A, Devlieghere F, Bernaerts K, Van Impe JF, Debevere J (2007) Growth/no growth models describing the influence of pH, lactic and acetic acid on lactic acid bacteria developed to determine the stability of acidified sauces. Int J Food Microbiol 119:258–269. doi:10.1016/j.ijfoodmicro.2007.08.003

Voit EO (2002) Models-of-data and models-of-processes in the post-genomic era. Math Biosci 180:263–274. doi:10.1016/S0025-5564(02)00115-3

Vorst KL, Todd ECD, Ryser ET (2006) Transfer of *Listeria monocytogenes* during slicing of turkey breast, bologna, and salami with simulated kitchen knives. J Food Prot 69:2939–2946

Vose D (2000) Risk analysis: a quantitative guide. Wiley, New York

Voysey P (2000) An introduction to the practice of microbiological risk assessment for the food industry applications. Guideline No 28. Campden and Chorleywood Food, Research Association Group, Chipping, Camden

Walsh S, Diamond D (1995) Non-linear curve fitting using Excel. Talanta 42:561–572. doi:0039-9140(95)01446-2

Whiting RC (1993) Modeling bacterial survival in unfavourable environments. J Ind Microbiol 12:240–246. doi:10.1007/BF01584196

Whiting RC, Buchanan RL (1994a) Microbial modeling. Food Technol 48:113–120

Whiting RC, Buchanan RL (1994b) Microbial modeling. IFT scientific status summary. Food Technol 44:110–119

Whiting RC, Cygnarowicz-Provost M (1992) A quantitative model for bacterial growth and decline. Food Microbiol 9:269–277. doi:10.1016/0740-0020(92)80036-4

Whyte W (1986) Sterility assurance and models for assessing air-borne bacterial contamination. J Parent Sci Technol 40:188–197

Wijtzes T, McClure PJ, Zwietering MH, Roberts TA (1993) Modelling bacterial growth of *Listeria monocytogenes* as a function of water activity, pH and temperature. Int J Food Microbiol 18:139–149. doi:10.1016/0168-1605(93)90218-6

Willocx F, Mercier M, Hendrickx M, Tobback P (1993) Modelling the influence of temperature and carbon dioxide upon the growth of *Pseudomonas fluorescens*. Food Microbiol 10:159–173. doi:10.1006/fmic.1993.1016

Wilson PDG, Brocklehurst TF, Arino S, Thuault D, Jakobsen M, Lange M, Farkas J, Wimpenny WT, Van Impe JF (2002) Modelling microbial growth in structured foods: towards a unified approach. Int J Food Microbiol 73:275–289. doi:10.1016/S0168-1605(01)00660-2

Wolff S, Antelmann H, Albrecht D, Becher D, Bernhardt J, Bron S, Bütner K, van Dijl JM, Eymann C, Otto A, Tam LT, Hecker M (2006) Towards the entire proteome of the model bacterium *Bacillus subtilis* by gel-based and gel-free approaches. J Chromatogr B 849:129–140

WTO (World Trade Organization) (1995) The WTO agreement on the application of sanitary and phytosanitary measures (SPS Agreement). http:/www.wto.org/english/trtop. Accessed 18 Feb 2012

Wu Y, Griffiths MW, McKellar RC (2000) A comparison of the Bioscreen method and microscopy for the determination of lag times of individual cells of *Listeria monocytogenes*. Lett Appl Microbiol 30:468–472. doi:10.1046/j.1472-765x.2000.00748.x

Zurera G, García-Gimeno RM, Rodríguez-Pérez MR, Hervás C (2004) Performance of response surface model for prediction of *Leuconostoc mesenteroides* growth parameters under different experimental conditions. Food Cont 17:429–438. doi:10.1016/j.foodcont.2005.02.003

Zwietering M (2005) Practical considerations on food safety objetives. Food Cont 16:817–823. doi:10.1016/j.foodcont.2004.10.022

Zwietering MH, Jongenburger I, Rombouts FM, Van't Riet D (1990) Modelling of the bacterial growth curve. App Environ Microbiol 56:1876–1881

Zwietering MH, Witjzes T, de Wit JC, Van't Riet K (1992) A decision support system for prediction of the microbial spoilage in foods. J Ind Microbiol 12:324–329. doi:10.1007/BF01584209

Zwietering MH, de Wit JC, Cuppers HG, van't Riet K (1994) Modeling of bacterial growth with shifts in temperature. Appl Environ Microbiol 60:204–213

Zwietering MH, de Wit JC, Notermans S (1996) Application of predictive microbiology to estimate the number of *Bacillus cereus* in pasteurised milk at the point of consumption. Int J Food Microbiol 30:55–70. doi:10.1016/0168-1605(96)00991-9

Index

F. Pérez-Rodríguez and A. Valero, *Predictive Microbiology in Foods*,   127
SpringerBriefs in Food, Health, and Nutrition 5, DOI 10.1007/978-1-4614-5520-2,
© Fernando Pérez-Rodríguez and Antonio Valero 2013

Printed by Printforce, the Netherlands